SUPERVISORY CONTROL OF DISCRETE EVENT SYSTEMS USING PETRI NETS

THE KLUWER INTERNATIONAL SERIES ON DISCRETE EVENT DYNAMIC SYSTEMS

Series Editor

Yu-Chi Ho
Harvard University

SUPERVISORY CONTROL OF DISCRETE EVENT SYSTEMS USING PETRI NETS

by

John O. Moody
Panos J. Antsaklis
University of Notre Dame
Notre Dame, IN USA

KLUWER ACADEMIC PUBLISHERS
Boston / Dordrecht / London

Distributors for North, Central and South America:
Kluwer Academic Publishers
101 Philip Drive
Assinippi Park
Norwell, Massachusetts 02061 USA

Distributors for all other countries:
Kluwer Academic Publishers
Distribution Centre
Post Office Box 322
3300 AH Dordrecht, THE NETHERLANDS

Library of Congress Cataloging-in-Publication Data

A C.I.P. Catalogue record for this book is available
from the Library of Congress.

*The publisher offers discounts on this book when ordered in bulk quantities. For
more information contact: Sales Department, Kluwer Academic Publishers,
101 Philip Drive, Assinippi Park, Norwell, MA 02061*

Printed on acid-free paper.

Printed in the United States of America

for my father,
Peter R. Moody, Jr.
(JOM)

to my teachers
to my students
(PJA)

Contents

List of Figures

List of Tables

Foreword

This book is one of a series of books in discrete event systems published by Kluwer since 1990. The subject matter has certainly grown since that time. It is actively covered in major conferences by INFORMS and IEEE. Looking back on the past decade, and at the foreword written for the first book in the series, we find the goals *to promote the study and understanding of the modeling, analysis, control, and management of Discrete Event Dynamic Systems* remain valid. The sub-goal of the modeling of DEDS has not changed either. Finite State Machines or Automata, Petri-nets (PN), and Generalized Semi-Markov Processes (GSMP) as representatives of simulation models still represent the three major contenders for general systems modeling (it can be argued that queuing networks, while very useful, are not general enough to model very complex and general systems).

A major computational difficulty with DEDS modeling is the combinatorial explosion of the state space. Since the state space is usually devoid of structural information or not easy to incorporate such information into it, this explosion often creates insurmountable computational burdens. In this sense, Petri-nets enjoy a distinction of directly embodying the structural information in its graphical model. Of course this is done at a price since there are limits to how large a graphical model the human mind can visualize at a glance. Hierarchical models of PN or colored PN are efforts at ameliorating such difficulties. As far as the analysis and control aspects of PN are concerned, this book develops the parallel of supervisory control for PN as has been done for finite state machines and automata. Thus, it is combining the advantages of both models. The authors' preface delineates in detail how this is done. We welcome this addition to the series.

Yu-Chi Ho, Consulting Editor
Harvard University, USA

Preface

This book is intended for graduate students, advanced undergraduates, and practicing engineers who are interested in the control problems of manufacturing, communication and computer networks, chemical process plants, and other high level control applications. The text is written from an engineering perspective, but it is also appropriate for students of computer science, applied mathematics, or economics. The book contains enough background material to stand alone as an introduction to supervisory control with Petri nets, but it may also be used as a supplemental text in a course on discrete event systems or intelligent autonomous control.

The book presents a novel approach for the supervisory control of discrete event systems using Petri nets. The concepts of supervisory control and discrete event systems are explained, and the background material on general Petri net theory necessary for using the book's control techniques is provided. A large number of examples are used to illustrate the concepts and techniques presented in the text, and there are plenty of references for those interested in additional study or more information on a particular topic.

Though the book contains a fair amount of explanatory and background material, some mathematical background is required. The reader should be familiar with basic concepts of matrix algebra and understand concepts such as null spaces or the kernel of an integer matrix. Fundamental ideas from the areas of mathematical programming and constrained optimization are also used here. The reader should be familiar with basic formulations of linear programming and with slack variables. Though the reader need not have encountered Petri nets before, this book is not intended as an introductory text on net theory. Some readers may wish to learn more about general net theory in parallel with or before learning the specific techniques covered here.

In the area of supervisory control, Petri nets provide an alternative to automata for modeling discrete event system plants. Petri nets provide increased modeling expressiveness and complexity compared to automata, allowing for a richer range of possible plant behaviors. Unfortunately this increase in modeling power typically also increases the difficulty of synthesizing supervisory controllers. This difficulty is overcome here for a large class of supervisory control goals using a method that is based on Petri net place invariants. The approach is easy to use and computationally efficient, both in the design and implementation phases, making the technique appropriate also for online control reconfiguration.

Following the introductory and background material of chapter 1, chapter 2 introduces the Petri net model for discrete event systems. The basic supervisor synthesis technique is described in chapter 3. Uncontrollable and unobservable transitions present important problems for supervisory control. These structures are introduced in chapter 4, and chapter 5 presents computational techniques for designing controllers in the face of these transitions. Chapters 6 and 7 describe an array of supervisory control problems that can be solved with the techniques of the previous chapters. These include the management of finite resources, deadlock avoidance, exclusions between events and states, and many others. A number of detailed examples are collected together in chapter 8 to illustrate the use and utility of the control techniques.

Chapter 2 contains all background material on Petri nets necessary to understand and use the supervisory control techniques of this book. The dynamic system model is described from both the graph theoretical and algebraic view points. Important concepts such as Petri net invariants, traps, and siphons are covered in detail. The different classes of Petri nets, based on their structural and thus behavioral properties, are described, and the relationship between Petri nets and automata is discussed. The chapter concludes with a brief survey of the uses of Petri nets as plant or controller models in supervisory control.

Chapter 3 presents the basic control technique of the book. Given a Petri net modeled plant and a linear inequality indicating a set of desired or safe plant states, a supervisor is constructed to insure that all reachable plant states conform to this inequality. The supervisory is itself a Petri net, connected to the transitions of the plant. The control design is very efficient, with the size of the supervisor being proportional to the number of constraints being enforced, and the computation of the controller's structure involving little more than a matrix multiplication. The controller is shown to be "maximally permissive," meaning that it will only prevent an event from occurring in the plant when that event would cause a violation of one or more of the constraints.

Actuator failures, practical concerns, or cost factors may make it impossible for a supervisor to prevent certain events from occurring during the plant's evolution. Similarly, sensor failures or other concerns may make it impossible for the controller to directly observe the occurrence of certain events. These ideas are translated into the concepts of uncontrollable and unobservable transitions in chapter 4. It is shown how individual plant constraints can be classified as admissible or inadmissible due to the presence of uncontrollable and unobservable transitions.

Chapter 5 presents computational techniques for transforming inadmissible constraints into admissible constraints. Once a given constraint has been made admissible using these techniques, the new transformed constraint can be enforced on the plant using the procedure of chapter 3. Techniques are also presented in chapter 5 for characterizing all admissible linear constraints on a plant with uncontrollable and unobservable transitions. This characterization can then be used to identify constraints that satisfy the requirements of an inadmissible constraint, but are themselves admissible. A controller is then constructed that enforces the disjunction of all these admissible constraints on the Petri net plant.

Linear inequalities placed on the reachable state space of a plant are useful for realizing a wide range of supervisory control specifications. Chapters 6 and 7 present a number of problems, some of which the reader might not have realized could be expressed using state-based inequalities. Chapter 6 discusses the modeling, management, and allocation of finite resources and presents supervisory techniques for liveness insurance and deadlock avoidance, problems that are often associated with the sharing of finite resources. Chapter 7 details a number of other control specifications, including the incorporation of events as well as states into constraints. Methods are presented for enforcing a class of logical predicates on system behavior, as well as means of including real time in the control specifications. Of course not all supervisory control specifications can be expressed as linear inequalities on the plant state, and chapter 7 includes a discussion on the limits of the control technique as well.

Small examples appear throughout the text to illustrate specific points and techniques as they are developed. Chapter 8 is used to collect the larger examples that incorporate multiple control procedures. This chapter need not be read in sequence. Readers may refer to specific examples within the chapter as they are cross referenced in the developmental material. The examples in chapter 8 include plants from the areas of flexible manufacturing, communications, process control, and hybrid systems.

Though the book follows a logical course, describing a particular approach to supervisory control, it is also intended to serve as a reference for readers who have become familiar with the material. Algorithms and mathematical propositions are easily identified within the text. The larger examples are cross referenced and collected within a single chapter for easy reference.

The end of the book contains a number of resources for the reader. All of the citations have been collected in a single section to make them easy to find while the book is being read and afterwards as a reference bibliography. The glossary covers many of the basics of Petri nets as well as the particular supervisory control techniques described in the text. A list of symbols is provided for the reader who wishes to understand a specific topic without going through all the developmental material. Finally, the complete index includes cross-references and topic sub-headings, making it easy to quickly find a specific subject in the text.

Readers of this book will learn the basics of supervisory control as well as an effective technique for its implementation using Petri nets. The reader will also receive an introduction to Petri nets, particularly the algebraic approach employed by the authors, showing that Petri nets can provide a useful paradigm for analysis, an efficient model for computation, and a powerful tool for implementation of supervisors for discrete event systems.

Acknowledgments

The authors gratefully acknowledge the support of the National Science Foundation, the Electric Power Research Institute, and the Army Research Office via grants NSF/MSS-9216559, EPRI/RP-8030-06, ARO/DAAH04-96-1-0134, ARO/DAAH04-95-1-0600, and NSF/ECS-9531485; the Arthur J. Schmitt Foundation and the University of Notre Dame's Center for Applied Mathematics for the fellowships they awarded; and of Notre Dame's Department of Electrical Engineering for providing an environment conducive to learning and academic pursuits.

JOHN MOODY AND PANOS ANTSAKLIS
University of Notre Dame, USA

1 INTRODUCTION

Discrete event systems (DES) are dynamic system models with state changes driven by the occurrence of individual events. The state space of a DES is a possibly infinite discrete set. State-to-state transitions are forced by the ordered occurrence of events from a discrete and (almost always) finite set. The discrete nature of the state and event spaces make these systems practical models for the higher level behavior of complex systems, capturing the logic and dynamics of overall system behavior and evolution. Applications for discrete event systems include industrial manufacturing systems, the relation between robotic actions and goals, process control plants, computer networks, communications protocols, and others.

This book deals with the automatic control of discrete event systems. It is often necessary to regulate or supervise the behavior of these systems in order to meet safety or performance criteria, e.g., preventing automated guided vehicles from colliding on a factory floor by restricting their access to certain mutually traveled zones. DES supervisors are used to insure that the behavior of the plant (system that is to be controlled) does not violate a set of constraints under a variety of operating conditions. The regulatory actions of the supervisor are based on observations of the plant state, resulting in feedback control.

It is common to see discrete event systems modeled as finite automata [Carroll, 1989, Wonham and Ramadge, 1987, Ramadge and Wonham, 1989]. Methods exist for designing controllers based on automata system models, however these methods often involve exhaustive searches or simulations of system behavior, making them impractical for systems with large numbers of states and transition causing events.

One way of dealing with these problems is to model discrete event systems with Petri nets (PN's). Petri nets [Peterson, 1981,Reisig, 1985,Murata, 1989] have a simple mathematical representation employing linear matrix algebra making them particularly useful for analysis and design. Petri net models are normally more compact than similar automata based models and are better suited for the representation of systems with ordered structures and flows but large reachable state spaces. The Petri net model allows for the simultaneous occurrence of multiple events, without suffering from increased model complexity, as is the case with automata. In addition they have an appealing graphical representation that has made them popular with practicing engineers. It is possible to visualize the state-flow of a system and to quickly see dependencies of one part of a system on another when the system is represented as a Petri net.

A Petri net is made up of places, which hold tokens, transitions, and directed arcs between the places and transitions. The number and distribution of tokens among the places is called the marking and represents the state of the net. The arcs determine how tokens move from one place to another upon the firing of a transition. The state of the system can be represented as an integer vector, and the architecture or layout of a Petri net can be represented with an integer matrix known as the incidence matrix.

The intuitive graphical representation and the powerful algebraic formulation of Petri nets has lead to their use in a number of fields. Petri nets are used to model multiprocessor computer systems, computer networks, digital communication protocols, process control plants, queuing systems, and flexible manufacturing cells, among others. Often times the graphical representation of a plant as a Petri net model is enough for an engineer to design a controller for the plant. Many control techniques exist that involve recognizing and then manipulating certain structures and situations that commonly appear in Petri net models. Other techniques exist for automatically verifying the reliability of these control designs. Representing the controller itself as a Petri net makes the verification of the combined plant/controller system simpler and reduces the number of computational tools required to model the overall system. Unfortunately, even when the controller is modeled as a Petri net, this cyclic technique of design and verification can become quite cumbersome when the plant model is large. This leads to the desire for an automatic method of generating controllers based on the plant and constraint data.

Through study of the structural invariants of Petri nets, which are actually elements of the kernel of the Petri net incidence matrix, and an application of some of the elementary concepts of mathematical programming, a method [Moody et al., 1994, Yamalidou et al., 1996] has been derived for synthesizing a supervisor for a Petri net based on the plant's model and a set of linear constraints to be imposed on the plant's behavior. The computation of the supervisor, or monitor [Giua et al., 1992], is simple and direct, making it very attractive compared to the searches employed in automata-based control synthesis procedures.

Other researchers have used Petri net structural invariants in their work on discrete event system control. [Valette et al., 1985, Valette, 1986] have explored the use of Petri nets as analytical and computational models for the representation of discrete event system controllers and have used Petri net place invariants as an analysis tool.

[Lautenbach and Ridder, 1994] use transition invariants as a means of examining and enforcing "liveness" on a Petri net. [Barkaoui and Abdallah, 1995, Huang et al., 1995, Tacconi et al., 1996, Huang et al., 1996] all make use of place invariants as a tool in the prevention of "deadlock" (see chapter 6). In this book, invariants are used to synthesize controls that enforce a set of specifications on the Petri net modeled plant. These specifications indicate allowed and forbidden states in the plant's state space.

The invariant-based control method requires that the constraints be linear inequalities on the Petri net marking vector. Fortunately it is possible to transform many other constraints on a plant's behavior into such inequalities, for example, [Yamalidou and Kantor, 1991] have shown how constraints written as Boolean expressions can be transformed into sets of linear inequalities involving the firing and marking vectors. Constraints involving the firing vector can be transformed, using two different methods, into constraints that involve only marking vector elements. Thus the control method can be applied to systems whose constraints can be expressed as linear inequalities or logic expressions and may involve elements of the marking and/or the firing vector.

The controller derived using this approach is maximally permissive in that it forces the set of constraints to be obeyed, while allowing any action that is not directly forbidden by the constraints.

A major goal in the field of discrete event system control is the synthesis of supervisors under conditions where certain state to state transitions can not be prevented by any action from the supervisor, i.e., conditions under which certain transitions are uncontrollable. The problem is then to design a controller that prevents states from occurring that violate the behavioral constraints directly or that might lead to a violation of the constraints through the action of uncontrollable transitions. A method for solving this problem using linear integer programming appears in [Li and Wonham, 1994]. A new method for solving this problem by performing algebraic manipulations on a matrix composed of the uncontrollable portion of the plant and a representation of the control goal has been proposed in several articles [Moody and Antsaklis, 1996a, Moody et al., 1995b, Moody et al., 1995a] These techniques are presented here and are expanded into a consistent treatment that offers both design elegance and computationally efficiency.

As in classical control theory, the concept of uncontrollability is associated with the dual concept of unobservability. It is possible that a DES plant might contain certain state to state transitions that can not be detected by the supervisor. The mathematical representation of these unobservable events is analogous to uncontrollable transitions [Moody and Antsaklis, 1997a, Moody and Antsaklis, 1996a, Moody et al., 1995a]. Both uncontrollable and unobservable transitions are covered by the design procedures of this book.

Chapter 2 gives an introduction to the Petri net model for discrete event systems. It includes a brief overview of related work in the area of DES control using Petri nets. Chapter 3 describes the invariant based controller, the basic supervisor design used throughout the book. Uncontrollable and unobservable transitions are introduced into the technique in chapter 4. Chapter 5 then presents computational techniques and algorithms for synthesizing controllers for plants with these transitions. Chapters 6 and 7 involve applications of the invariant based control technique to common DES supervisory control problems. Finite resource management and deadlock avoidance are covered in chapter 6. Chapter 7 examines control specifications that are not normally expressed with respect to the plant's state but can still be tackled with the invariant based control technique. These specifications include constraints on allowed events, logic-based behavior constraints, and constraints involving real time. Chapter 8 contains a number of detailed examples that illustrate the ideas discussed in the preceding chapters. This chapter may be read in sequence, or the reader may wish to refer to the individual sections of the chapter when they are referenced in the developmental material of chapters 3 through 7.

2 PETRI NETS

Petri nets and finite state machine (FSM) automata lack the full modeling and decision power that can be implemented with microprocessor based digital computers, which have the complete expressiveness of the Turing machine. In terms of formal languages, Petri net languages include all regular languages, but not all context-sensitive (or even context-free) languages [Peterson, 1981]. Specifics regarding these details are discussed in section 2.5.1. It is logical to then ask why Petri nets are such a popular modeling paradigm for discrete event systems when they lack the full expressiveness of the Turing machine. A degree of modeling power is sacrificed in order to take advantage of the simple and direct graphical representation of Petri nets and, more importantly, the elegant and efficient mathematical representation that makes them easy to code and invites the use of many well known mathematical analytical tools. These reasons combined with the fact that many practical systems can be effectively modeled with PN languages make the Petri net an excellent choice for many DES modeling and control tasks.

This chapter provides a brief introduction to Petri nets and their use in DES control. Fundamental PN definitions are covered in section 2.1. The structural invariants of a Petri net are very important to the work of this book and are covered separately in section 2.2. Section 2.3 covers the concepts of Petri net traps and siphons, while section 2.4 presents the various classes of Petri nets based on their structural properties. The material of these two sections is relevant to the discussion of deadlock, deadlock avoidance, and liveness in chapter 6. The relation of Petri nets to automata, and automata based supervisory control, is covered in section 2.5. The chapter concludes in section 2.6 with a brief overview of the work of other researchers in using Petri nets as a DES control tool.

2.1   Petri Net Definitions

A Petri net is a directed bipartite graph. The structure of a Petri net is described by (P, T, D^+, D^-) where P and T are disjoint sets representing the vertices of the graph, known as *places* and *transitions*, and D^+ and D^- are integer matrices with nonnegative elements representing the flow relation between the two vertex types.

A place is a storage cell for *tokens*. When a transition *fires*, a number of tokens are removed from some places and added to others. An *arc* with weight D_{ij}^+ from transition j to place i indicates that when transition j fires, place i will receive D_{ij}^+ tokens. An arc with weight D_{kj}^- from place k to transition j indicates that place k must contain at least D_{kj}^- tokens before transition j is allowed to fire and that when transition j fires, place k will lose D_{kj}^- tokens. Thus for a transition to fire, all of its *input places* must contain a minimum number of tokens. A transition that meets these conditions is *enabled* and is free to fire. A *disabled* transition may not fire. When a transition fires, all of its input places lose a number of tokens, and all of its *output places* gain a number of tokens.

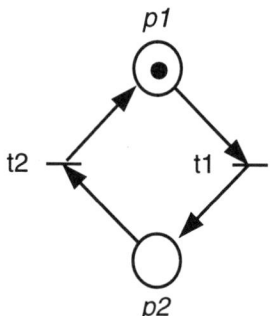

Figure 2.1.   A simple Petri net.

Example.   Figure 2.1 shows a simple Petri net; p_1 and p_2 are the places, t_1 and t_2 are the transitions, and the arcs are represented with arrows. If an arc in a drawing is not assigned a weight then it is assumed the weight is one; all four arcs in Figure 2.1 have a weight of one. Tokens are represented as dots; place p_1 contains the net's single token in Figure 2.1. Transition t_1 is the only enabled transition. When it fires, its input place, p_1, will lose a token, and its output place, p_2, will gain a token, at which time t_2 will be the only enabled transition.

Let $\mathbb{Z}$ be the set of integers and let n be the number of places in a Petri net and m be the number of transitions. The arcs connecting transitions to places are described by the matrix $D^+ \in \mathbb{Z}^{n \times m}$ and the arcs connecting places to transitions are described by $D^- \in \mathbb{Z}^{n \times m}$ where all of the elements of D^+ and D^- are greater than or equal to zero.

The distribution of tokens throughout the net is called the Petri net *state* or *marking*. The marking is represented with the n dimensional integer vector μ. The *initial marking* of the net is μ_0. The transitions of a Petri net fire in discrete steps. The transitions that are to fire at the current step are represented by the m dimensional integer vector q. The jth element of q is 0 if the jth transition will not be firing, and the jth element of q is 1 if the jth transition will fire. A given *firing vector q* represents a valid possible firing if all of the transitions for which it contains nonzero entries are enabled. The validity of a given firing vector q can be determined by checking the *enabling condition*:

$$\mu \geq D^- q \qquad (2.1)$$

Unless otherwise stated, *all vector and matrix inequalities in this book are read element-by-element with respect to the two sides of the inequality,* in this case the vectors μ and $D^- q$. Thus the enabling condition can be thought of as the logical conjunction of n inequalities.

The Petri net *incidence matrix* is defined as

$$D = D^+ - D^- \qquad (2.2)$$

If the given Petri net contains no *self-loops*, i.e. no transition has input and output arcs involving the same place, then the enabling condition can be expressed in terms of the complete incidence matrix. Without self loops, a particular transition firing can either add or subtract tokens from a given place, but it can not do both. Thus the non zero places of the D^+ and D^- matrices are mutually exclusive. Because all elements of $D^+ q$ must, by definition, be greater than or equal to zero and because of the mutual exclusivity of the nonzero elements in D^+ and D^-, we can use the following inequality as an equivalent enabling condition to (2.1).

$$\begin{aligned}
\mu + D^+ q &\geq D^- q \\
\mu + (D^+ - D^-)q &\geq 0 \\
\mu + Dq &\geq 0
\end{aligned} \qquad (2.3)$$

When a Petri net contains no self-loops, the D matrix is uniquely defined by D^+ and D^-, and since (2.3) can be used rather than (2.1) for the enabling condition, the D^+ and D^- matrices are not normally specified. It is not necessary that a plant have no self-loops to use the synthesis tools of this book, however the self-loop free notation will often be used for convenience.

Care must be taken when using (2.1) or (2.3) when q indicates the concurrent firing of multiple transitions. There are a variety of different techniques for handling concurrency. Concurrency may not be allowed at all, in which case q would be a zero vector with a single element equal to one. Concurrency may be allowed only when each of the indicated transition firings could occur one after the other in any order. In this case q must satisfy (2.1), or each transition firing indicated in q must independently satisfy (2.3) as well as the complete q vector. If (2.3) is used without this check for independently enabled transitions, then certain concurrent firings may be allowed even though some or all of the individual transitions indicated in the firing could not fire by themselves. The choice of which of these methods to use is dictated by the modeling requirements and the particular plant.

When the transitions described by q fire, the state of the Petri net will change. The state change is described by

$$\mu \Leftarrow \mu + Dq \tag{2.4}$$

Note that the left hand side of (2.3) appears in (2.4): the state vector of a Petri net will never contain a negative element. Thus when no self-loops are present, a Petri net can be described by the following system,

$$\begin{aligned} \mu(0) &= \mu_0 \\ \mu(k+1) &= \mu(k) + Dq(k) \end{aligned} \tag{2.5}$$

$$D \in \mathbb{Z}^{n \times m}, \mu \in \mathbb{Z}^n, q \in \mathbb{Z}^m, (\mu, q \geq 0) \tag{2.6}$$

which is similar to a discrete-time linear system with constraints placed on the state and input vectors. The PN enabling condition is a consequence of the constraint that the state and input vectors contain no negative elements.

The concepts of existence and uniqueness of solutions for difference equations can not be applied to system (2.5) due to the constraints placed on the state and input vectors by (2.6). These constraints, that μ and q can be composed only of nonnegative integers, are a cause of difficulty and complexity in Petri net analysis, but they are also the reason for the utility of the systems as DES models. A similar situation occurs in the area of mathematical programming where an optimization problem is defined over an entire vector space or module, but only constrained elements are considered feasible solutions. Without (2.6), system (2.5) is a rather uninteresting subclass of linear difference equations, but with (2.6), it becomes something entirely different, requiring different tools and concepts for analysis.

A net is called *safe* if no place in the net can ever contain more than a single token. The majority of work in this book does not rely on a net being safe or even *bounded*, but safe nets are used in section 7.3 and will be mentioned on other occasion.

Example.   The Petri net of Figure 2.1 has the incidence matrix

$$D = D^+ - D^- = \begin{bmatrix} 0 & 1 \\ 1 & 0 \end{bmatrix} - \begin{bmatrix} 1 & 0 \\ 0 & 1 \end{bmatrix} = \begin{bmatrix} -1 & 1 \\ 1 & -1 \end{bmatrix}$$

and contains no self loops. The current state is

$$\mu = \begin{bmatrix} 1 \\ 0 \end{bmatrix}$$

The only enabled firing is, according to the enabling condition (2.3),

$$q = \begin{bmatrix} 1 \\ 0 \end{bmatrix}$$

After the firing defined by q occurs, the state will change according to (2.4):

$$\underbrace{\begin{bmatrix} 0 \\ 1 \end{bmatrix}}_{\mu} \Leftarrow \underbrace{\begin{bmatrix} 1 \\ 0 \end{bmatrix}}_{\mu} + \underbrace{\begin{bmatrix} -1 & 1 \\ 1 & -1 \end{bmatrix}}_{D} \underbrace{\begin{bmatrix} 1 \\ 0 \end{bmatrix}}_{q}$$

Inspection of Figure 2.1 shows that this is a safe net with a single token that bounces back and forth between the two input places.

2.2   Structural Invariants

One of the *structural properties* of Petri nets, i.e. properties that depend only on the topological structure of the Petri net and not on the net's initial marking, are the net invariants. Invariants are important means for analyzing Petri nets since they allow for the net's structure to be investigated independently of any dynamic process [Lautenbach, 1987].

Place invariants correspond to sets of places whose weighted token count remains constant for all possible markings. They are represented by n-dimensional integer vectors x, where n is the number of places of the Petri net; non-zero entries correspond to the places that belong to the particular invariant. A place invariant is defined as every integer vector $x \in \mathbb{Z}^n$ that satisfies

$$x^T \mu = x^T \mu_0 \tag{2.7}$$

where μ_0 is the net's initial marking, and μ represents any subsequent marking. Equation (2.7) means that the weighted sum of the tokens in the places of the invariant remains constant for all reachable markings, and this sum is determined by the initial of the Petri net. The place invariants of a net can be computed by finding integer solutions to

$$x^T D = 0 \tag{2.8}$$

where D is the $n \times m$ incidence matrix of the Petri net. Given any firing vector q,

$$\begin{aligned} x^T \mu(k+1) &= x^T(\mu(k) + Dq(k)) \\ &= x^T \mu(k) \end{aligned}$$

It is easily shown that any linear combination of place invariants is also a place invariant for the net. Place invariants are a primary analytical tool used in the control synthesis work of this book.

The dual of the place invariant is the *transition invariant*. A transition invariant $y \in \mathbb{Z}^m$ satisfies

$$Dy = 0$$

Using the Petri net system definition (2.5) we see that

$$\begin{aligned} \mu(0) &= \mu_0 \\ \mu(1) &= \mu_0 + Dq(0) \\ \mu(2) &= \mu(1) + Dq(1) \\ &= \mu_0 + Dq(0) + Dq(1) \\ &\vdots \\ \mu(N) &= \mu_0 + D(q(0) + \cdots + q(N-1)) \\ &= \mu_0 + DQ \end{aligned} \tag{2.9}$$

where Q is the sum of N firing vectors. Equation (2.9) indicates that *the state of a Petri net after N firings is independent of the order in which the firings occur.* Thus if $Q = y$, i.e., a transition invariant, then

$$\mu(N) = \mu_0 + Dy = \mu_0$$

Transition invariants represent a set of firings that will cause the marking of a net to cycle, leaving it in the state that it held before the start of the cycle. *The existence of a transition invariant does not imply that it will actually be possible to fire the indicated transitions*; the initial conditions of a net may prohibit it. Instead a transition invariant indicates that if it is possible to fire the given set of transitions, in any order, the state of the net will return to its initial condition at the end of the sequence.

2.3   Siphons and Traps

Traps and siphons[1] (see [Desel and Esparza, 1995, Murata, 1989, Reisig, 1985]) are sets of Petri net places. *Once the set of places in a trap become marked, the trap will always be marked for all future reachable markings. Similarly, once the marking of a siphon becomes empty, the siphon will remain empty.* These structures are fundamental to the analysis of liveness and for creating supervisors that perform deadlock-avoidance (see chapter 6).

Traps and siphons are defined by the nature of the input and output transitions into a given set of places. Let $\bullet p$ refer to the set of input transitions into the place p, and let $p\bullet$ refer to the set of output transitions from the place p. The "bullet" notation can also be used with sets of places. If S is a set of places, then $\bullet S$ and $S\bullet$ refer to the set of input and output transitions for the entire set S.

Definition 2.1 A set of places S is a **siphon** iff

$$\bullet S \subseteq S\bullet$$

S is a **minimal siphon** iff there does not exist another siphon P such that $P \subset S$.   ■

The definition of a trap is similar.

Definition 2.2 A set of places S is a **trap** iff

$$S\bullet \subseteq \bullet S$$

S is a **minimal trap** iff there does not exist another trap P such that $P \subset S$.   ■

A place invariant vector with nonnegative elements indicates a set of places that is both a trap and siphon. The net of Figure 2.2a shows why the invariant must be nonnegative. This net contains the invariant

$$\mu_1 - \mu_2 = 0$$

---

[1] Siphons are sometimes called "deadlocks."

but the two places are neither a trap nor a siphon.

Given that a set of places is both a trap and an invariant, it is not necessarily true that the set also represents the support of a place invariant. For example, consider the net of Figure 2.2b. The two places form both a trap and a siphon, however this net contains no invariants.

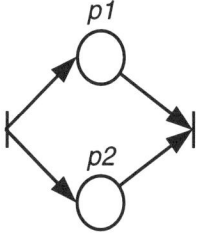

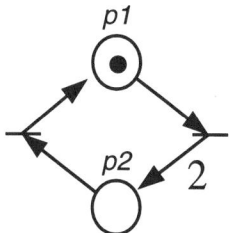

Figure 2.2a.   Invariant with no trap or siphon.

Figure 2.2b.   Trap and siphon with no invariant.

Algorithms for computing the traps or siphons of a Petri net appear in [Ezpeleta et al., 1993] and [Lautenbach, 1987]. The work of [Ezpeleta et al., 1993] includes a comparison of several siphon-calculation techniques as well as their own algorithm, which is presented below.

Definition 2.3 Let $x \in \mathbb{Z}^n$ be an integer vector corresponding to the places of a Petri net. The **support** of x, $\|x\|$, is the set of places corresponding to the nonzero entries in x. ∎

Remark. If x represents a place invariant vector, then $\|x\|$ is the set of places that forms the invariant. Similarly, $\|\mu\|$ is the set of all marked places. Traps and siphons are sets, but computational techniques for finding these sets involve integer vectors. The notion of support is used to go between the two representations.

Proposition 2.4 Characterization of siphons. Given a PN with incidence matrix D, if $x^T D \leq 0$, $x \geq 0$, then $S = \|x\|$ is a siphon in the Petri net.

Proof. See [Ezpeleta et al., 1993]. ∎

Remark. If x is a place invariant, then $x^T D = 0$. The quantity $x^T D$ is often referred to as the *defect* when $\|x\|$ is a siphon but not a place invariant.

The technique for calculating siphons involves examining the structure of a transformed version of the original Petri net. The following proposition provides a necessary and sufficient condition for relating siphons in one PN (D') to another (D).

Proposition 2.5 Necessary and sufficient siphon characterization. Given two Petri nets with incidence matrices $D = D^+ - D^-$ and $D' = D'^+ - D'^-$ and

$$D'^+ (p, t) = 0 \quad \leftrightarrow \quad D^+ (p, t) = 0$$
$$D'^- (p, t) = 0 \quad \leftrightarrow \quad D^- (p, t) = 0$$

where $D(p, t)$ is the element of D corresponding to place p and transition t. If

$$\forall t, \forall p_i \in \bullet t, D'^-(p, t) \geq \sum_{p' \in t \bullet} D'^+(p', t)$$

then $S = \|x\|$, $x \geq 0$, is a siphon in the PN with incidence matrix D iff

$$x^T D' \leq 0 \tag{2.10}$$

Proof. See [Ezpeleta et al., 1993].   ∎

All siphons in D can be found by calculating values of x that meet inequality (2.10). The transformed incidence matrix D' can be computed as follows:

$$D'^+ = D^+$$

$$\forall t, \forall p \in \bullet t, D'^-(p, t) = kD^-(p, t)$$

where

$$k = \sum_{p' \in t \bullet} D^+(p', t)$$

and $D' = D'^+ - D'^-$.

Inequality (2.10) can be changed into an equality through the addition of slack variables:

$$\begin{bmatrix} x^T & z^T \end{bmatrix} \begin{bmatrix} D' \\ I \end{bmatrix} = 0 \tag{2.11}$$

where $x, z \geq 0$. Equation 2.11 has the form of a PN place invariant equation. Thus appropriate values of x can be found by finding the set of minimal support place invariants of the PN with incidence matrix $\begin{bmatrix} D' \\ I \end{bmatrix}$.

An algorithm for computing minimal support place invariants appears in [Martinez and Silva, 1980]. The results obtained will yield a *generating family* of siphons for the original Petri net. All siphons in a net can be found as unions of the members of the generating family. A generating family will include all of the net's minimal siphons, however the set of minimal siphons will not, in general, include every member of the generating family. For example, the minimal siphon of the Petri net of Figure 2.3 is $\{p_1\}$, but the generating family is $\{\{p_1\}, \{p_1, p_2\}\}$.

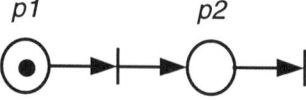

p1   p2

Figure 2.3.   The generating family of siphons is larger than the set of minimal siphons.

Calculating the generating family of traps for a Petri net with incidence matrix D is accomplished by calculating the generating family of siphons for a net with incidence matrix $-D$.

2.4   Classes of Petri Nets

Petri nets are classified based on structural properties involving the ways arcs connect places and transitions. Some results in the Petri net literature are only applicable for certain classes of nets. Nearly all of the results in this book apply to general nets with no assumptions made on their structural properties, however some of the results in chapter 6 on deadlock and liveness apply only to *asymmetric choice* or *extended free choice* Petri nets. These and the other important classes of Petri nets (see [Desel and Esparza, 1995, Murata, 1989]) are described below.

Definition 2.6 A **state machine (SM)** or **S-system**[2] is a Petri net in which all transitions have one input and one output place.   ∎

The significant nodes in a state machine are the places. Each transition allows tokens to flow from one place to another, but a token in a particular place may enable multiple transitions. This is referred to as a *conflict*. All finite automata can be described as Petri net state machines.

Definition 2.7 A **marked graph (MG)** or **T-system** is a Petri net in which all places have a single input and a single output transition.   ∎

The significant nodes in a marked graph are the transitions. Each place receives tokens from one transition and loses tokens to another, but a single transition may have multiple input and output places. Marked graphs allow *synchronization,* since tokens in multiple places are simultaneously lost due to the firing of a single output transition and gained by the firing of a single input transitions. Marked graphs contain no conflicts and state machines have no synchronizations. Some nets, e.g., a closed loop of places and transitions, have neither conflict nor synchronization and are both marked graphs and state machines. *Free choice* nets allow both conflicts and synchronizations.

Definition 2.8 A **free choice (FC)** net is a Petri net such that for every arc from a place p to a transition t, $p \to t$,

1. t is the only output transition of p (no conflict), or

2. p is the only input place of t (no synchronization).

∎

All state machines and marked graphs fall under the class of free choice nets. The definition of free choice nets shows that if any output transition of a place p is enabled, then all output transitions of p are enabled. Most work on free choice nets uses a slightly weaker definition to extend the class of nets while maintaining this basic property.

---

[2] The 'S' in S-system refers to the German word *stellen*, which means "place." Petri's dissertation on net theory was written in German, and his notation is still used.

Definition 2.9 An **extended free choice (EFC)** net is a Petri net such that for every arc $p \rightarrow t$ there exists an arc from all input places of t to all all output transitions of p.
∎

Using the "bullet" notation, for all pairs of places, p_1 and p_2, in an EFC net,

$$\text{if } p_1 \bullet \cap p_2 \bullet \neq \emptyset, \text{ then } p_1 \bullet = p_2 \bullet$$

The structural restrictions are relaxed even more for *asymmetric choice* (AC) nets.

Definition 2.10 An **asymmetric choice (AC)** net is a Petri net such that for all pairs of places, p_1 and p_2,

$$\text{if } p_1 \bullet \cap p_2 \bullet \neq \emptyset, \text{ then } p_1 \bullet \subseteq p_2 \bullet \text{ or } p_2 \bullet \subseteq p_1 \bullet$$

∎

The Venn diagram of Figure 2.4 illustrates the hierarchy of the PN classes and provides examples of nets that fall in the various categories.

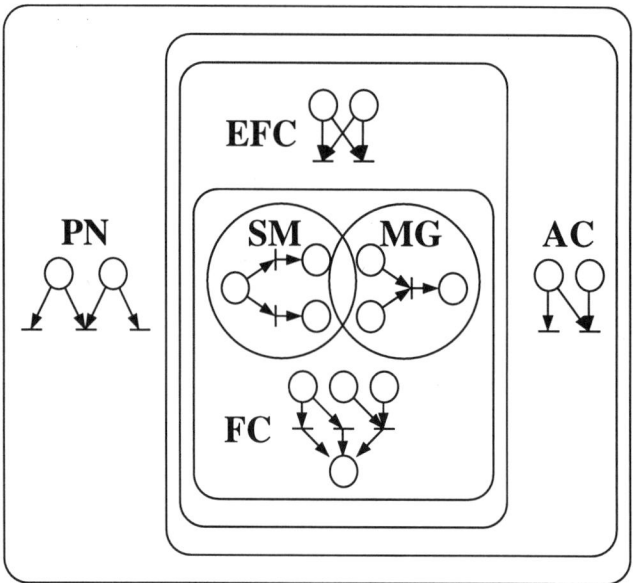

Figure 2.4.   Venn diagram showing the relation of PN classes.

2.5   Petri Nets and Automata

2.5.1   Formal Languages

The modeling and representational power of various automata can be expressed in terms of the type of language the given automata is able to "speak." A list of the different types of languages and the associated automata are described here as presented by [Carroll, 1989] and [Révész, 1983].

Finite automata (both deterministic and nondeterministic) speak *type 3 (regular) languages*, the least powerful of the formal languages. Finite automata receive a set of symbols and react to these symbols in the sequence that they are received.

"Pushdown automata" or finite automata with stacks are capable of understanding context-free *type 2 languages*. Structured high level computer programming languages are type 2 languages, thus compilers for these languages can be described as pushdown automata. Type 2 languages make it possible to handle such things as matched parentheses and *begin / end* pairs in programming languages.

The most advanced of the formal languages, types 1 and 0, are accepted by Turing machines and two-pushdown automata. When the size of these machines is bounded, they accept the context-sensitive *type 1 languages* of linear bounded automata, otherwise they speak the more general *type 0 (phrase structure) languages*. The Turing machine is capable of randomly accessing its stream of input signals, thus allowing for loops and many of the similar powerful features associated with modern digital computers. In fact, any sequential digital computer can (in theory) be modeled as a Turing machine. A summary of these formal languages and their associated automaton models is given in Table 2.1.

Table 2.1.   The formal languages and their automaton models.

	Name	Automaton model
Type 3	Regular	Finite state machines
Type 2	Context-free	Pushdown automata
Type 1	Context-sensitive	Linear bounded automata
Type 0	Phrase structure	Turing machines, Two-pushdown automata

Ordinary Petri nets do not speak the most expressive of the formal languages, but they do have advantages over standard finite automata. As an example of the increased efficiency enjoyed by Petri net models over finite automata, consider the language

$$a^k b a^l b \qquad\qquad 0 \le l \le k \qquad\qquad (2.12)$$

where a and b are the symbols of the language. In a Petri net, the occurrence of a language symbol can be represented by the firing of a transition. A Petri net that realizes language (2.12) is shown in Figure 2.5 (see [Giua and DiCesare, 1994]). There are an infinite number of possible states for this Petri net, but the Petri net model only requires a four place, four transition graph. Representing language (2.12) with an automaton would require an infinite graph, as shown in Figure 2.6.

Language (2.12) is context-free, but it is not regular. A Petri net can be constructed so that the firing of its transitions represent any regular language [Peterson, 1981], but clearly, as seen in Figures 2.5 and 2.6, not all Petri net languages are regular.

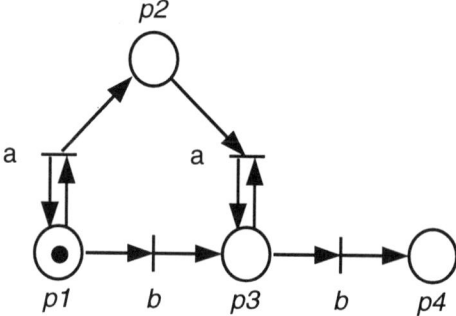

Figure 2.5.   This infinite state Petri net is represented with a finite graph. It accepts a context-free language that is not regular.

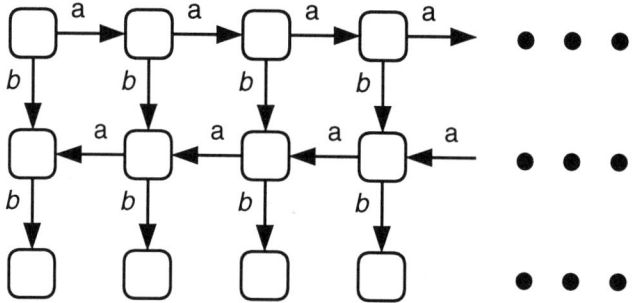

Figure 2.6.   An infinite state automaton with an infinite graph. The context-free language can not be represented with a finite automaton.

Not all context-free languages are Petri net languages either [Peterson, 1981]. Suppose we modify language (2.12) such that we force the number of occurrences of a, before and after the first occurrence of b, to be equal, i.e.,

$$a^k b a^k b \qquad\qquad k \geq 0 \qquad\qquad (2.13)$$

In this case, the language remains context-free, but it is no longer possible to construct an ordinary Petri net that accepts this and only this language [Giua and DiCesare, 1994]. If there is a bound placed on k, then a Petri net can be constructed for (2.13), as shown in section 2.5.2. Language (2.13) can be realized by a Petri net if the PN model is expanded to include the use of *inhibitor arcs*[3]" [Giua and DiCesare, 1994], but then we are no longer dealing with ordinary Petri nets.

Just as not all context-free languages are Petri net languages, not all Petri net languages are context-free either. The network in Figure 2.7 accepts the following

---

[3] An *inhibitor arc* is an arc $p \rightarrow t$ that prevents t from firing when p contains a number of tokens greater than or equal to the arc weight.

language

$$a^k b^l c^m \qquad\qquad 1 \leq m \leq l \leq k \qquad\qquad (2.14)$$

which is context-sensitive, but not context-free [Peterson, 1981].

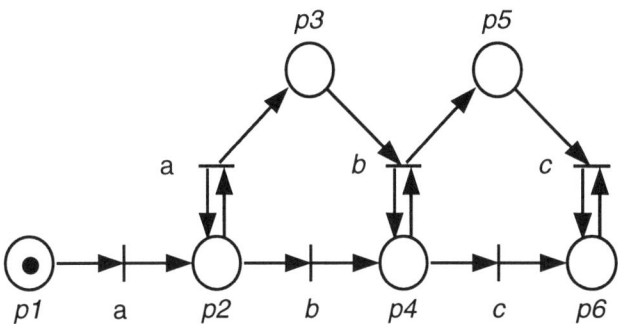

Figure 2.7. A Petri net for language (2.14).

[Peterson, 1981] provides a useful explanation for understanding the relationship between the PN languages and the context-free and context-sensitive languages. The size of the state space of a pushdown automaton, speaking a context-free (type-2) language, grows exponentially with the length of the input string. However, the state space of a Petri net grows only combinatorially with the length of the input string. Clearly, the larger state space of the pushdown automata allows them to represent languages that can not be represented by Petri nets. However, at any given time, a pushdown automaton has access only to the current input and the top element of its stack. A Petri net, on the other hand, with its comparatively complex interconnections between places and transitions, has access to a large number of counters at any given time. This allows the Petri net to represent languages that can not be represented by a pushdown automaton, despite the smaller state space size.

A summary of the PN language's relation to the other formal languages appears in the Venn diagram of Figure 2.8.

2.5.2 DES Control

Automata models may not always have the most efficient representations, but they do lend themselves to the design of supervisors that prevent forbidden or undesirable states from occurring in an automaton modeled plant. Consider the discrete event system represented as an automaton in Figure 2.9a and as a Petri net in 2.9b. Our control goal is to make sure that we never visit state (0,3) (the shaded box in Figure 2.9a) in the automaton. In terms of the Petri net, this means that we wish to insure that the number of tokens in place two, μ_2, remains less than or equal to three, i.e. we wish to create a controller that enforces the linear constraint $\mu_2 \leq 3$.

It is possible to construct an automaton supervisor to solve this problem by first duplicating the structure in 2.9a. The next step is to strip out the undesirable state, and then to assign to each of the remaining states the plant transitions that are allowed to occur. The controller is shown in Figure 2.10. The controller is used by initializing it

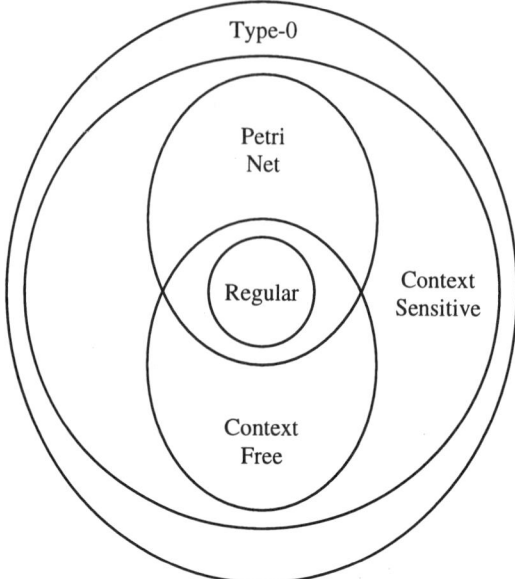

Figure 2.8.   The set of Petri net languages includes all regular languages, intersects the context-free languages, and is a subset of the context-sensitive languages.

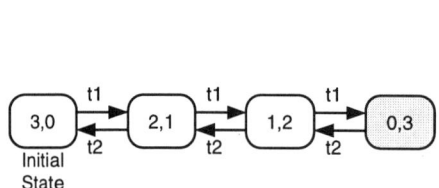

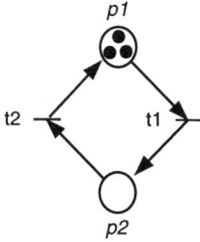

Figure 2.9a.   Automaton model of a simple discrete event system.

Figure 2.9b.   Petri net representation of the system.

to the analogous starting state in the plant. The current state of the controller indicates which of the plant transitions it will permit to occur. When the plant does advance to a new state, the transition is noted by the controller, and the controller advances its own state accordingly. At this time the controller has a new set of transitions that it will allow to occur. The process then continues.

Control design using Petri net models is not quite as simple. It is not possible to duplicate the Petri net structure and remove the forbidden states because a Petri net graph does not contain a unique structure for every state. Instead of attempting to use supervisory control to prohibit certain prespecified forbidden states, the approach taken in this book is to use supervisors to enforce linear predicates on the state vectors

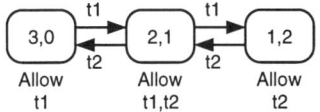

Figure 2.10.   The automaton controller for the system of Figure 2.9a

of Petri nets. Enforcement of these types of constraints can be used to solve a variety of supervisory control problems (see chapter 7) as illustrated in the example below.

Example.   Given the Petri net of Figure 2.5, we wish to use supervisory control to insure that the net only speaks languages of the form

$$a^k b a^k b \qquad\qquad 0 \le k \le K \qquad\qquad (2.15)$$

where K is a predefined constant. Note that this language is the bounded version of (2.13). Language (2.15) is a subset of (2.12), the language of the unsupervised Petri net of Figure 2.5. Supervisory control can be used to restrict (2.12) such that all strings of transition firings for the network follow the form given in (2.15).

An examination of Figure 2.5 indicates that enforcing language (2.15) means insuring that the deadlocking transfer of a token into p_4 is not allowed to occur until after p_2 has been completely emptied of tokens. Language (2.15) indicates that p_2 may contain at most K tokens, so restricting language (2.12) to only allow (2.15) is equivalent to forcing the following constraint on the state space of the Petri net:

$$\mu_2 + K\mu_4 \le K \qquad\qquad (2.16)$$

Chapter 3 shows how a supervisor for enforcing a constraint in the form of (2.16) can be generated automatically. The supervised Petri net is shown in Figure 2.11 with $K = 3$. The figure is drawn to emphasize that the original plant model of Figure 2.5 has not been modified. The supervisor is a feedback device that places limits on the open loop behavior of the plant. The closed loop system does not include new behaviors not present in the open loop, unsupervised plant.

2.6   Petri Nets in Control

Many researchers have used Petri nets as a tool for modeling, analyzing and synthesizing control laws for discrete event systems. [Murata et al., 1986] defined C-nets as an extended form of safe Petri nets and used them to construct station controllers for sequencing control with quick response time. [Boissel, 1993] used simulated annealing to compute a Petri net controller for a discrete-event system modeled by a Petri net. [Zhou and DiCesare, 1989] proposed the adaptive design of Petri net controllers for automated manufacturing systems. They define the controller as the control logic based on the Petri net model of the process and use the model to generate a supervisory controller by successive augmentation. [Boucher and Jafari, 1992] have presented a method of transforming controller designs using the "Structured Analysis and Design Technique" and the "Integrated Computer Manufacturing Definition 0" into Petri nets in order to take advantage of the their graphical and computational efficiencies.

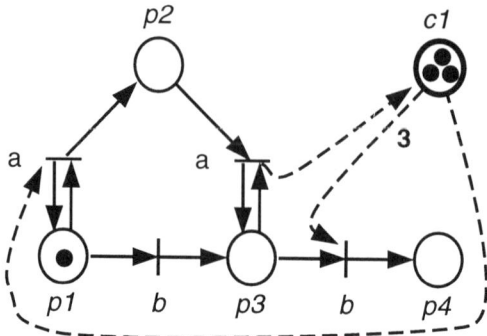

Figure 2.11. A supervisor is connected to the Petri net of Figure 2.5 such that the language of the closed loop system is given by (2.15) where $K = 3$.

Many problems in the field of chemical engineering process control lend themselves to discrete event and Petri net representations. A survey of these issues appears in [Cohen et al., 1985] and includes Petri net descriptions and their use as a modeling and analysis tool for process control problems. [Yamalidou and Kantor, 1991] develop means of creating modular process control model components and logical behaviors using Petri nets.

The mathematical descriptions of the structural properties of a Petri can be used to formulate and solve optimization problems for control and planning. Proth et al. have formulated such problems for Petri nets that have timing requirements associated with the transition firings. [Hervé and Proth, 1989] used deterministically-timed Petri net representations of manufacturing processes. They developed a heuristic algorithm for solving a integer linear programming problem that determines the minimum distribution of jobs in progress that leads to a full utilization of bottleneck (limiting) machines. This work is extended by [Laftit et al., 1992] to show how the initial conditions of a timed Petri net can be chosen such that the constant weighted sum of a place invariant is minimized[4] while maintaining a given bound on the net's cycle time. This works lends itself, for example, to the problem of maximizing the productivity of a job shop in a manufacturing system while maintaining a minimal work-in-progress level. Further analysis of these systems appears in [Cohen et al., 1985], and the work of [Laftit et al., 1992] was extended by [Proth and Xie, 1994] to apply to timed nets that have stochastic variations in the firing times.

Other researchers have also associated stochastic processes with the firing times of Petri nets or with the flow of queuing networks modeled as Petri nets. [Ross, 1986] investigated a procedure for deciding where inputs to a set of queues should be routed based on a cost function that accounts for the average number of units waiting in a queue. [Guo et al., 1993] used moment generating functions to associate transfer

---

[4] Recall that a place invariant is a weighted sum of markings that remains constant for all reachable markings within a Petri net. The actual value of the weighted sum is determined by the initial conditions of the net.

functions with transitions that have a probability of firing (like a Markov process) and a time delay (which can be deterministic or stochastic). The method was used to find a number of performance measures such as cycling time, mean sojourn time, mean recurrence time, steady state probabilities, as well as some transient performance parameters.

As in [Guo et al., 1993], other researchers have successfully applied tools tradition-ally associated with continuous time systems to the field of DES. [Passino et al., 1994] demonstrate how Lyapunov functions can be associated with discrete event dynamic systems, including Petri nets, to prove certain stability and performance properties of these systems.

[Sreenivas and Krogh, 1992] have investigated some of the implications of the infinite state space of Petri net models and the implications to supervisory control. In this work, the researchers have examined nets that include inhibitor arcs, which increase the modeling power over that of ordinary Petri nets but, as a tradeoff, can decrease the computational tractability of the model as a control tool.

[Giua et al., 1992] developed an efficient technique for transforming Petri net behav-ioral constraints into Petri net modeled controllers. This method produces controllers, or monitors, identical to the ones discussed in chapter 3 of this book. More recently these researchers have done work on language control and realization with Petri nets [Giua and DiCesare, 1994], extending some of the control synthesis ideas for automata developed in [Wonham and Ramadge, 1987, Ramadge and Wonham, 1989].

[Holloway and Krogh, 1990] developed a method to control systems that can be modeled as cyclic marked graphs, a special class of Petri nets. They have shown how to synthesize a maximally permissive feedback controller that guarantees to prevent a set of forbidden states from occurring. Their method does not require an exhaustive search of the system's state space, is computationally effective and polynomial in the number of forbidden conditions and in the number of places of the net. The drawback is that it is applicable to a limited class of systems. These authors present an informative study of other Petri net control issues in [Holloway and Krogh, 1994] and [Holloway et al., 1997].

[Li and Wonham, 1993, Li and Wonham, 1994, Li and Wonham, 1995] have extended the work of [Wonham and Ramadge, 1987, Ramadge and Wonham, 1989], on DES supervisory control, to vector DES's (Petri nets). The problem they address deals primarily with the handling of uncontrollable transitions within the plant structure. They show how, under certain assumptions about the plant structure, a control law can be found to enforce a set of linear constraints by solving a linear integer programming problem. The work of Li and Wonham is important to the topic of this book and is discussed in more detail in section 4.4.

The controller synthesis method presented here computes a controller based on the plant's place invariants, a structural property of Petri nets. The controller consists of places that are connected to the transitions of a plant insuring that the state vector of the plant remains within bounds established by a set of linear constraints. The combined plant/controller net possesses the necessary place invariants to insure that the set of constraints is not violated. The controller is easily computed (involving only matrix multiplications) and its size is proportional to the number of constraints of the plant. The controller is maximally permissive in that it forces the set of constraints to be obeyed, while allowing any action that is not directly forbidden by the constraints. Analysis of this basic procedure then leads to methods for enforcing constrained behavior while accounting for uncontrollable and unobservable transitions in the plant.

3 INVARIANT BASED CONTROL DESIGN

Consider the supervisory control goal of restricting the reachable markings of a plant, μ_p, such that

$$l^T \mu_p \leq b \tag{3.1}$$

where l is an integer weight vector, and b is an integer scalar. Constraints of the form of (3.1) are not as general as the language based behavioral constraints of [Wonham and Ramadge, 1987, Ramadge and Wonham, 1989] used for automaton based models. Linear state constraints are used here because of the ease with which they can be enforced on Petri net modeled systems and because, though they lack some of the full power of language based constraints, they are still quite versatile. Clearly these linear constraints on the state space can represent a large number of (convex) forbidden state problems. They are also useful for realizing "generalized mutual exclusion constraints" [Giua et al., 1992], which includes both serial and parallel mutual exclusions [Desrochers and Al-Jaar, 1995]. Many constraints that deal with exclusions between events and states, or just events themselves, can be transformed into constraints with the form of (3.1). These linear inequalities also lend themselves to constraints dealing with the use or sharing of finite resources and can in some instances even be used to model constraints that deal with absolute and relative timing requirements. Section 5.4 shows how a slight modification of the controller behavioral rules can allow the enforcement of nonconvex constraints formed as the disjunction of multiple linear inequalities. Chapter 7 deals with the varied uses of these linear inequalities on the plants state; the chapter comes late in the book because it makes use of many of the ideas and synthesis procedures that are developed in the chapters that precede it.

The method presented in this chapter is a powerful means of realizing these constraints because it is simple to calculate, and the Petri net structure of the solution makes the controller easy to implement. The supervisor structure, designed using the concept of Petri net place invariants, was derived by [Moody et al., 1994] and [Yamalidou et al., 1996] and produces controllers identical to the monitors of [Giua et al., 1992]. The approach has a number of advantages:

1. The design method is transparent for theory, analysis, and implementation as it is based on the concept of place invariants.

2. The resulting controller and consequently the overall controlled system are described by simple Petri nets.

   (a) A variety of tools for analysis both graphical and algebraic are available; verification of the design is therefore rather straightforward.

   (b) The automatic handling of concurrent events is maintained with a unified plant/controller Petri net model.

3. The design method has excellent numerical properties that make it particularly appealing for large problems or control reconfiguration applications where, because of a failure, the controller must be redesigned on-line.

The system to be controlled is modeled by a Petri net with n places and m transitions and is known as the *plant* or *process net*. The incidence matrix of the plant is $D_p \in \mathbb{Z}^{n \times m}$. It is possible that the plant will violate certain constraints placed on its behavior, thus the need for control. The *controller net* is a Petri net with incidence matrix D_c made up of the plant's transitions and a separate set of places. The *controlled net* is the Petri net with incidence matrix D made up of both the original plant and the added controller. The controlled net is also called the *controlled system* or the *closed loop system*.

The control goal is to force the plant to obey constraints of the form (3.1). For example, we might wish to enforce the constraint $\mu_1 + \mu_2 \leq 1$, which means that at most one of the two places p_1 and p_2 can be marked, or, in other words, both places cannot be marked at the same time, and neither place can ever have more than a single token.

This inequality constraint can be transformed into an equality by introducing a nonnegative *slack variable* μ_c into it. The constraint then becomes $\mu_1 + \mu_2 + \mu_c = 1$ or, in general

$$l^T \mu_p + \mu_c = b \tag{3.2}$$

The slack variable in this case represents a new place c that holds the extra tokens required to meet the equality. The rules of Petri net state evolution insure that μ_c is, by definition, nonnegative. So if equality (3.2) can be forced on the marking of the controlled system, the weighted sum of tokens in the plant's places will always be less than or equal to b.

The place that enforces the inequality constraint is part of a separate net called the controller net. The structure of the controller net will be computed by observing that the introduction of the slack variable introduces a place invariant,[1] defined by equation (3.2), for the overall controlled system.   Section 3.1 provides an overview of the *monitors* of Giua and DiCesare due to their relation to the invariant based controllers of section 3.2. The method is shown to be maximally permissive in section 3.3.

3.1   Monitor Based Supervisors

This section provides a summary and discussion of the Petri net control concept of monitors introduced by [Giua et al., 1992]. The topic is discussed here because the monitors themselves are identical to the place invariant controllers discussed in section 3.2. Monitors are used to enforce *generalized mutual exclusion constraints* (GMEC's), which are defined by the set

$$\mathcal{M}(l, b) = \{\mu_p \in \mathbb{Z}^n, \mu_p \geq 0 | l^T \mu_p \leq b\}$$

where n is the number of places, and μ_p is the marking of the plant. A group of GMEC's can be written as

$$
\begin{aligned}
\mathcal{M}(L, b) &= \bigcap_{i=1}^{n_c} \mathcal{M}(l_i, b_i) \\
&= \{\mu_p \in \mathbb{Z}^n, \mu_p \geq 0 | L\mu_p \leq b\}
\end{aligned}
\tag{3.3}
$$

where b is now a vector.

A GMEC is *redundant* with respect to the marking $\mu_p \in \mathbb{Z}^n$ if $\mu_p \subseteq \mathcal{M}(l, b)$. A GMEC is redundant to the Petri net N with initial marking μ_{p0}, $\langle N, \mu_{p0} \rangle$, if $R(N, \mu_{p0}) \subseteq \mathcal{M}(l, b)$ where $R(N, \mu_{p0})$ is the set of reachable markings of N starting with marking μ_{p0}.

Proposition 3.1   Redundant GMEC's. If the following linear program (LP) has an optimal solution $x^* < b + 1$ then the GMEC, $\mathcal{M}(l, b)$, is redundant with respect to $\langle N, \mu_{p0} \rangle$.

$$z = \max l^T \mu_p$$
$$\text{s.t.} \quad X\mu_p - X\mu_{p0}$$
$$\mu_p \geq 0$$

where X is the basis of the *P-semiflows* of the net.

Proof. See [Giua et al., 1992].   ∎

Remark. The "P-semiflows" refer to the Petri net place invariants. It is not difficult to discuss redundancy in light of invariant theory.

Demonstrating the equivalence of two GMEC's is more complicated than simply demonstrating that the two constraints are linearly dependent. Net structure plays an

---

[1] A brief review of Petri net place invariants appears in section 2.2.

important role in determining case by case equivalence. Two sets of GMEC's (L_1, b_1) and (L_2, b_2) are *equivalent* with respect to $\langle N, \mu_{p0} \rangle$ if $R(N, \mu_{p0}) \cap \mathcal{M}(L_1, b_1) = R(N, \mu_{p0}) \cap \mathcal{M}(L_2, b_2)$. Linear programs for checking this condition are presented in [Giua et al., 1992].

The equivalence of GMEC's leads to the idea of the possible simplification of constraints. GMEC (l_1, b_1) is *simpler* than (l_2, b_2) if $l_1 < l_2$. The simplest constraints are vectors composed of 1's and 0's and are called "unweighted" constraints. Theorem 1 in [Giua et al., 1992] shows that a weighted GMEC can always be reduced to a set of unweighted GMEC's if the net is safe.

GMEC's are not general enough to realize all forbidden state problems. This is demonstrated with a simple example. Let $\mu_{p1}, \mu_{p2}, \mu_{p3} \in R(N, \mu_{p0})$ be reachable markings with $\mu_{p3} = \frac{1}{2}(\mu_{p1} + \mu_{p2})$, then

$$\mu_{p1}, \mu_{p2} \in \mathcal{M}(L, b) \rightarrow \mu_{p3} \in \mathcal{M}(L, b)$$

because $\mathcal{M}(L, b)$ is a convex set. However a general forbidden state problem might be to prohibit markings μ_{p1} and μ_{p2} while allowing μ_{p3}. Theorem 2 in [Giua et al., 1992] demonstrates that general forbidden state problems on safe and conservative (entire net is covered by a nonnegative place invariant) nets can always be represented with GMEC's.

GMEC's can be enforced on the marking behavior of a plant with *monitors* that are equivalent to the controllers presented in the following section. A monitor is itself a Petri net with its own set of places and arcs that are attached to the transitions of the plant. Each GMEC requires a controller place, equivalent to a slack place, and an initial condition. The equations are identical to those given in section 3.2. The only differences lie in how the equations are derived and/or justified. The following results for monitor based controllers appear in [Giua et al., 1992].

1. A monitor insures that the projection of the reachable set of the controlled system onto the uncontrolled system meets the given constraint.

2. The projection of the potentially reachable set of the controlled system is identical to the legal potentially reachable markings of the uncontrolled net.

3. The monitor minimally restricts the plant's behavior, i.e., it only prohibits transition firings that would induce forbidden markings.

4. Liveness is not necessarily guaranteed by monitor based controllers. Furthermore, given a particular initial marking, it may not be possible to reach all possible allowed markings, and it might not even be possible to return to the initial marking. These consequences are not a result of monitor based control but are due to the strict, literal enforcement of the constraints.[2]

There are some thoughts on the problems of enforcing constraints in the face of uncontrollable transitions in [Giua et al., 1992]. The authors note that, for safe and

---

[2] Means for handling liveness and deadlock issues appears in chapter 6.

conservative nets, control can still be imposed even in the face of uncontrollable transitions. [Li and Wonham, 1993, Li and Wonham, 1994] have addressed the solvability of maximally permissive control problems under conditions of uncontrollability for a wider range of PN models. Chapters 4 and 5 present an approach for a computationally more efficient means of dealing with uncontrollable and unobservable transitions.

3.2   Supervisor Synthesis using Place Invariants

Each constraint of type (3.1) enforced on the net has a slack variable associated with it and each slack variable is represented in the controller net as a place. Thus the size (number of places) of the controller net is proportional to the number of constraints that are to be enforced. Every place used to control the plant adds one row to the incidence matrix D of the controlled system. Thus D is composed of two matrices, the original D_p of the plant model and the incidence matrix of the controller, D_c. The arcs connecting the controller place to the original Petri net of the system are computed by the place invariant equation (2.8) where the unknowns are the elements of the new row of D and the vector x is the desired place invariant defined by equation (3.2), i.e. $x^T = [l_1\ l_2\ \dots\ l_n\ 1]$.

The control problem can be stated in general as follows. All constraints of type (3.1) can be grouped and written in matrix form as

$$L\mu_p \leq b \tag{3.4}$$

where μ_p is the marking vector of the plant, $L \in \mathbb{Z}^{n_c \times n}$, $b \in \mathbb{Z}^{n_c}$, and n_c is the number of constraints of type (3.1). The inequality is with respect to the individual elements of the two vectors $L\mu_p$ and b and can be thought of as the logical conjunction of n_c separate inequalities. Disjunctions of inequalities are discussed in section 5.4. Constraint (3.4) is equivalent to the grouped GMEC constraint (3.3) of [Giua et al., 1992].

After the introduction of the slack variables, constraint (3.4) becomes

$$L\mu_p + \mu_c = b \tag{3.5}$$

where $\mu_c \in \mathbb{Z}^{n_c}$ integer vector that represents the marking of the controller places.

The matrix D_c contains the arcs that connect the controller places to the transitions of the plant. The incidence matrix $D \in \mathbb{Z}^{(n+n_c) \times m}$ of the closed loop system is given by

$$D = \begin{bmatrix} D_p \\ D_c \end{bmatrix} \tag{3.6}$$

and the marking vector $\mu \in \mathbb{Z}^{n+n_c}$ and initial marking μ_0 are

$$\mu = \begin{bmatrix} \mu_p \\ \mu_c \end{bmatrix} \qquad \mu_0 = \begin{bmatrix} \mu_{po} \\ \mu_{co} \end{bmatrix} \tag{3.7}$$

Note that equation (3.5) is in the form of (2.7), thus the invariants defined by equation (3.5) on the system ((3.6), (3.7)) must satisfy equation (2.8) with x being replaced by X, a matrix specifying the n_c different invariants.

$$X^T D = [L \ I] \begin{bmatrix} D_p \\ D_c \end{bmatrix} = 0$$

$$LD_p + D_c = 0 \tag{3.8}$$

where $I \in \mathbb{Z}^{n_c \times n_c}$ is an identity matrix since the coefficients of the slack variables in equation (3.5) are all equal to 1.

Theorem 3.2 Controller synthesis. If

$$b - L\mu_{p_0} \geq 0 \tag{3.9}$$

then a Petri net controller, $D_c \in \mathbb{Z}^{n_c \times m}$ with initial marking μ_{c_0}

$$D_c = -LD_p \tag{3.10}$$

$$\mu_{c_0} = b - L\mu_{p_0} \tag{3.11}$$

enforces constraints (3.4) when included in the closed loop system (3.6) with marking (3.7), assuming that the transitions with input arcs from D_c are controllable.

If inequality (3.9) is not true, then the constraints can not be enforced since the initial conditions of the plant lie outside the range defined by the constraints.

Proof. If inequality (3.9) is not true, then obviously $L\mu_{p_0} > b$ and the initial conditions of the plant violate the constraints. If the inequality is true, then equation (3.11) shows that the initial conditions of the controller are defined as a vector representing the slack in each of the constraints represented by $L\mu_p \leq b$.

Equation (3.10) is the solution of equation (3.8), which forces $\begin{bmatrix} L & I \end{bmatrix}$ to be invariants of the closed loop system. Since $\begin{bmatrix} L & I \end{bmatrix}$ are invariants we know that $L\mu_p + \mu_c = L\mu_{p_0} + \mu_{c_0}$ and from equation (3.11), $L\mu_{p_0} + \mu_{c_0} = b$. The marking of a place is always greater than or equal to zero, therefore

$$L\mu_p + \mu_c = b$$
$$L\mu_p \leq b$$

∎

Remark. Theorem 3.2 creates a controller that will enable and inhibit various transitions in the plant. If any of these transitions are uncontrollable then the controller defined by this method is invalid. Chapter 5 shows how a transformation of the constraints can be performed so that the uncontrollable transitions in the net receive no input arcs from the controller places.

Remark. The method can control transitions that participate in self-loops in the plant, since no assumptions regarding self-loops were required in the theorem.

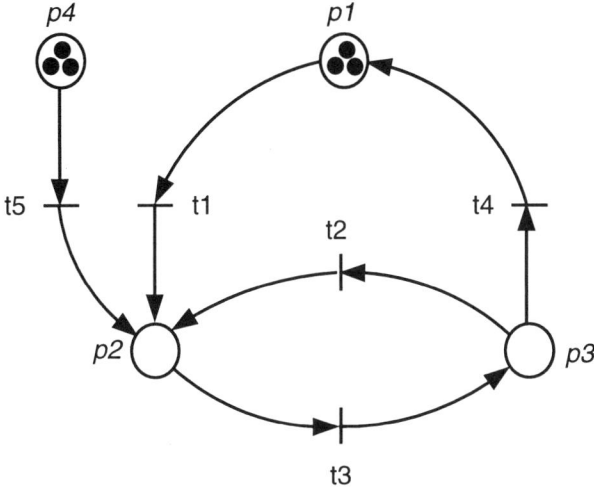

Figure 3.1.   Plant net for the example of section 3.2

Remark. When an element of D_c is zero, there are no arcs at all connecting the given place and transition, i.e., there are no cancelling self-loops in the controller structure. Self-loops may occur in the controller if the graph transformation techniques (for realization of firing vector constraints) discussed in section 7.2 are used.

Example.   Consider the Petri net of Figure 3.1, which is acyclic and non-safe. Its incidence matrix is

$$D_p = \begin{bmatrix} -1 & 0 & 0 & 1 & 0 \\ 1 & 1 & -1 & 0 & 1 \\ 0 & -1 & 1 & -1 & 0 \\ 0 & 0 & 0 & 0 & -1 \end{bmatrix}$$

and its initial marking is

$$\mu_{p_0} = \begin{bmatrix} \mu_1 \\ \mu_2 \\ \mu_3 \\ \mu_4 \end{bmatrix} = \begin{bmatrix} 3 \\ 0 \\ 0 \\ 3 \end{bmatrix}$$

D_p is rank 3, thus it has one place invariant that includes the entire net, i.e., $X^T D_p = 0$ where $X^T = [1\ 1\ 1\ 1]$. The objective is to control the net so that places p_2 and p_3 never contain more than one token, i.e. we wish to enforce the constraint

$$\mu_2 + \mu_3 \leq 1 \tag{3.12}$$

Using the matrix notation of (3.4) we have

$$L = \begin{bmatrix} 0 & 1 & 1 & 0 \end{bmatrix}$$

$$b = 1$$

The plant will not satisfy the desired constraint without an external controller. A slack variable μ_c is introduced and inequality (3.12) becomes an equality

$$\mu_2 + \mu_3 + \mu_c = 1 \tag{3.13}$$

The slack variable μ_c denotes the marking of a controller place, p_c. Equation (3.13) represents the desired invariant $X^T = [0\ 1\ 1\ 0\ 1]$ that will be forced on the closed loop system. The incidence matrix of the controller net is computed with equation (3.10)

$$D_c = -LD_p = \begin{bmatrix} -1 & 0 & 0 & 1 & -1 \end{bmatrix}$$

The initial marking of the controller place is computed from equation (3.11)

$$\mu_{c_0} = 1 - L\mu_{p_0} = 1$$

The Petri net graph of the controlled system is shown in Figure 3.2. The controller arcs are shown with dashed lines and the control place with thick lines.

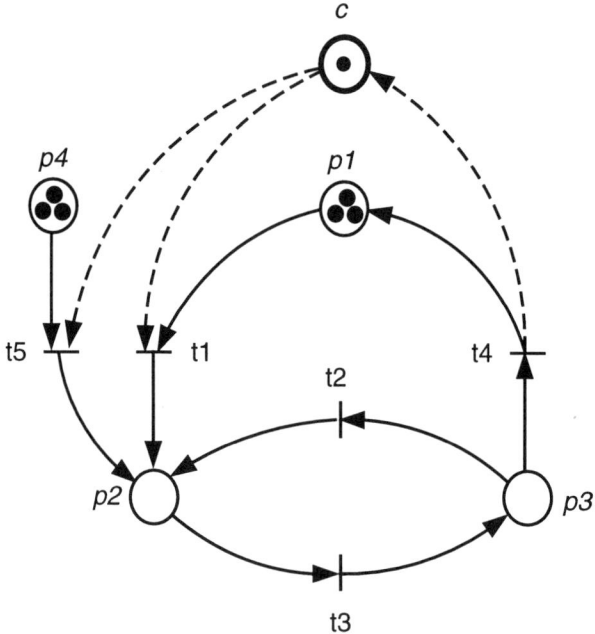

Figure 3.2.   The Petri Net of Figure 3.1 with Controller.

Transitions 1 and 5 are enabled, but place c prevents them from firing concurrently or from firing any time that a token occupies p_2 or p_3.

Further illustrations of the basic control synthesis procedure appear among the examples of chapter 8. Section 8.1.1 uses the "cat and mouse" problem and section 8.2 involves collision avoidance among "automated guided vehicles" on a factory floor.

3.3   Maximally Permissive Supervision

The invariant based control method is maximally permissive.[3] The Petri net enabling condition indicates that a transition is inhibited only if its firing would cause the marking of any of its input places to become negative. Thus a controller place only acts to inhibit a transition when the firing would cause $\mu_c < 0$. The invariant forced by the controller on the closed loop system, $l^T \mu_p + \mu_c = b$, shows that if μ_c is negative, then $l^T \mu_p > b$. The controller will only act to inhibit in situations where the firing of a transition would cause a direct violation of the constraint inequality.

Next it is shown, by examination of the place invariants of the controlled system, that the control scheme only introduces those invariants that are required by the constraints. Let X_p be an integer matrix of linearly independent columns representing a basis for the place invariants of the (uncontrolled) plant. Then X_p satisfies the following equation

$$X_p^T D_p = 0$$

where the number of columns of X_p (and thus the number of invariants) is equal to $(n - \text{rank } D_p)$ since D_p is an $n \times m$ matrix and X_p forms a basis for the kernel of D_p. Note that if rank $D_p = n$ then the uncontrolled plant has no place invariants.

A controller is constructed using equation (3.10). The incidence matrix of the controlled net is then

$$D = \begin{bmatrix} D_p \\ D_c \end{bmatrix} = \begin{bmatrix} D_p \\ -LD_p \end{bmatrix}$$

The rows of D_c are linear combinations of the rows of D_p, so rank $D = $ rank D_p. Thus the number of invariants of the controlled system is equal to $n + n_c - $ rank D_p. All of these invariants are accounted for by the uncontrolled plant invariants and the forced constraint invariants as shown below. First note that

$$\begin{bmatrix} X_p^T & 0 \end{bmatrix} \begin{bmatrix} D_p \\ D_c \end{bmatrix} = X_p^T D_p = 0$$

Thus the invariants of the uncontrolled plant are also invariants of the controlled plant. This is true for any Petri net control scheme that only adds places and arcs in order to control the plant. From the construction of the control law we also know that

$$\begin{bmatrix} L & I \end{bmatrix} \begin{bmatrix} D_p \\ D_c \end{bmatrix} = LD_p + D_c = LD_p - LD_p = 0$$

and thus all $n + n_c - \text{rank} D_p$ invariants of the controlled net are given by $X_c^T D = 0$ where

$$X_c = \begin{bmatrix} X_p & L \\ 0 & I \end{bmatrix}$$

---

[3] An exception to the general maximal permissivity of invariant based controllers may occur when constraints have the form of equalities rather than inequalities. This condition may occur inadvertently due to interaction between the constraint inequalities and the plant's natural invariants. Attempts to enforce equality constraints may lead to deadlock. See section 7.1 for details.

The rank (and number of columns) of X_c is $n + n_c - \text{rank}D_p$, since X_p is rank $n - \text{rank}D_p$ and I is a $n_c \times n_c$ identity matrix.

There are no new or unexpected invariants forced on the system as a result of the control law. The control law is maximally permissive since no action is prohibited that is not a result of the plant structure itself or a particular constraint forced on the plant.

4 UNCONTROLLABLE AND UNOBSERVABLE TRANSITIONS

Consider the situation where the controller is not allowed to influence certain transitions in the plant Petri net. It is illegal (in most cases[1]) for the Petri net controller to include an arc from one of the controller places to any of these *uncontrollable* plant transitions, because these arcs are the mechanism by which the controller exerts its power to inhibit events.

A major goal in the field of discrete event system control is the synthesis of supervisors under conditions where certain state to state transitions can not be prevented by any control action [Ramadge and Wonham, 1989, Wonham and Ramadge, 1987, Li and Wonham, 1994, Holloway and Krogh, 1990, Holloway et al., 1997, Sreenivas, 1997b]. For example, the plant Petri net may include models for failures or breakdowns, which obviously can not be controlled, or there may be models of irreversible processes, like many chemical reactions, which can not be stopped once started. For example, consider a constraint in a chemical processing plant that indicates that no more than five tanks should contain chemicals undergoing exothermic reactions at the same time. A control method that did not account for uncontrollable transitions might mix reactants in all of the tanks and then attempt to stop the actual reaction from taking place in enough of the tanks in order to meet the safety criterion. The problem is then to design a controller that prevents states from occurring that directly violate the

---

[1] A controller may contain arcs to an uncontrollable transition if the controller never acts to inhibit that transition when it is otherwise enabled. See section 4.1 for details. Corollary 4.8 provides a condition under which these kinds of arcs are permitted.

behavioral constraints or *that might lead to a violation of the constraints through the action of uncontrollable transitions.*

In the hypothetical chemical processing plant above, the uncontrollable transitions could be dealt with by restating the constraint so that it dealt with the mixing of the reactants as well as the number of tanks containing active reactions. However for large, complex systems it may be too difficult to foresee all of the conflicts between the constraints and the uncontrollable transitions.

It is also possible that certain plant transition may be *unobservable*, yet still modeled in the Petri net plant. These events may be too expensive or even impossible to monitor directly, or they may be the result of sensor failures.

In this chapter, the concepts of uncontrollable and unobservable transitions are defined and explained in sections 4.1 and 4.2. The architectural limitations these transitions place on a Petri net supervisor are fundamental results used in the analysis of section 4.5 and for the synthesis procedures of chapter 5. The use of constraint transformations to deal with uncontrollable and unobservable transitions is introduced in section 4.3 through the concepts of *admissible* and *inadmissible* markings and constraints. The research into "vector discrete event systems" by Li and Wonham includes important results relevant to the process of transforming constraints to account for uncontrollability. An overview of these results is presented in section 4.4. The mathematical preliminaries and tools used in the procedures of chapter 5 are presented in section 4.5 at the conclusion of this chapter.

4.1   Uncontrollable Transitions

Definition 4.1   A plant transition is called **uncontrollable** if the firing of that transition can not be inhibited by an external action.   ∎

The freedom of an uncontrollable transition to fire is limited solely by the structure and state of the plant. In order for a Petri net controller to inhibit a transition, it must contain an arc from a controller place to the transition. The transition will be disabled if the number of tokens in the control place is less than the arc weight. However this is not the only function of the arc between the two elements. When a plant transition fires, any control place connected to it will gain or lose tokens and experience a change in its marking: the firing causes the controller to undergo a state change. *For a Petri net modeled controller, the ability of the controller to inhibit a transition is intrinsically tied with its ability to change state based on observations of that transition.* Section 4.2 includes more information on this aspect of Petri net controllers in its discussion of unobservable transitions.

Definition 4.1 allows for an uncontrollable transition to be observed, thus arcs may be freely drawn from uncontrollable transitions to control places. In some situations, a Petri net control place may even contain an arc to an uncontrollable transition, but there must be assurance that if the number of tokens in the control place ever falls below the arc weight of the connection, then the transition is also simultaneously disabled by at least one of its input places in the plant. Corollary 4.8 in section 4.5 provides a situation where this condition is guaranteed, and the example of the "three tank problem" in section 8.6 provides a practical illustration of its use.

The synthesis techniques in chapter 5 will often make use of an incidence matrix corresponding to the uncontrollable portion(s) of the plant. Let D_{uc} be an incidence matrix composed of the columns of D_p that correspond to uncontrollable transitions. $D_{uc} \in \mathbb{Z}^{n \times n_{uc}}$ where n_{uc} is the number of uncontrollable transitions. Given a set of constraints, $L\mu_p \leq b$, the Petri net controller is given by $D_c = -LD_p$. Recall that, assuming no self loops, positive elements in an incidence matrix refer to arcs from transitions to places, and negative elements refer to arcs from places to transitions. So, ignoring the few situations where a controller can observe a transition without ever inhibiting it, the controller will violate the uncontrollability condition if LD_{uc} contains any elements greater than zero. That is, the uncontrollability of a transition indicates that we can not draw any arcs from the controller places to this transition, and the portion of the controller corresponding to the uncontrollable transitions is given by $-LD_{uc}$, thus the elements of LD_{uc} should be less than or equal to zero. This result is formally proposed in Corollary 4.7 of section 4.5.

4.2   Unobservable Transitions

It is possible that transitions within the plant may be unobservable, i.e., they are defined on the Petri net graph because they represent the occurrence of real events, but these events are either impossible or too expensive to detect directly. It is also possible, in the event of a sensor failure, that a transition might suddenly become unobservable, forcing a redesign or adaptation of the control law.

Definition 4.2 A plant transition is called **unobservable** if the firings of that transition can not be directly detected or measured.   ∎

Since the firing of an unobservable transition can not be detected, a controller state change can not be triggered by such a firing. For a Petri net based controller, both input and output arcs to the plant transitions are used to trigger state changes in the controller. *A Petri net controller can not have any connections to an unobservable transition, thus all unobservable transitions are also implicitly uncontrollable.* One can imagine a situation where the occurrence of some event in a plant could be blocked without the controller ever receiving any feedback relating directly to that event, but, in practical situations, the ability to inhibit an event is usually coupled with the ability to detect occurrences of that event. For this reason, the limitation on Petri net based controllers with regard to unobservable transitions is not too severe.

Similar to D_{uc}, let $D_{uo} \in \mathbb{Z}^{n \times n_{uo}}$ be an incidence matrix composed of the columns of D_p corresponding to unobservable transitions, where n_{uo} is the number of these transitions. The incidence matrix of a Petri net controller for the constraint $L\mu_p \leq b$ is given by $D_c = -LD_p$. Thus, in order for the controller to conform to the limitations imposed by unobservability, all of the elements of LD_{uo} must be equal to zero.

4.3   Constraint Transformations

Given a set of constraints, $L\mu_p \leq b$, a supervisor must work to insure that the constraints are never violated directly or may be violated through the firing of uncontrollable transitions or through incomplete knowledge due to unobservable transitions.

One approach to this problem would be to construct a supervisor that behaved according to the rules below. The procedure assumes that unobservable transitions are also uncontrollable.

At every iteration and for each enabled controllable transition in the net:

1. If the firing of the transition would lead to a state μ_p such that $L\mu_b \not\leq b$ (for any element in the vector-to-vector inequality), then disable this transition.

2. Else search for any markings that are reachable through the firing of uncontrollable transitions (from the state that would be induced by the firing of the candidate transition). If any of these reachable markings violates the constraints, then disable the transition.

3. Else the transition is enabled.

Unlike the controller structures discussed in chapter 3, a supervisor acting according to the rules above would not take the form of external Petri net modifying the behavior of the plant net. The primary disadvantage of this approach lies in the potentially expensive search for uncontrollably reachable markings, which leads to restrictions on the classes of plants that may be studied, or which may prevent the supervisor from being implemented in real time in an actual application.

A separate approach to supervising a plant with uncontrollable/unobservable transitions is to actually modify the constraints themselves such that the new constraints account for the difficult structures in the plant. If it is possible to obtain an analytic solution for the transformed constraints, then the controller logic itself will be very simple. The following definitions are useful in understanding the motivation for the transformation of constraints. The definitions are with respect to a plant with possible uncontrollable or unobservable transitions and a constraint on the marking behavior of the plant in the form $L\mu_p \leq b$. Unobservable transitions are also assumed to be uncontrollable.

Definition 4.3 An **admissible marking** μ_p is a marking such that

1. $L\mu_p \leq b$, and

2. For all markings μ_p' reachable from μ_p through the firing of uncontrollable transitions, $L\mu_p' \leq b$.

If either of these conditions is not met, then the marking is **inadmissible**.   ■

Definition 4.4 Given a plant with initial marking μ_{p_0}, an **admissible constraint** satisfies

1. $L\mu_{p_0} \leq b$, and

2. $\forall \mu_p$ reachable from μ_{p_0} such that $L\mu_p \leq b$, μ_p is an admissible marking.

If a constraint does not satisfy both of these conditions, then it is **inadmissible**.   ■

If a constraint is admissible[2], then condition 2 of Definition 4.4 indicates that the firing of uncontrollable transitions can never lead from a state that satisfies the constraint to a new state that does not satisfy the constraint.

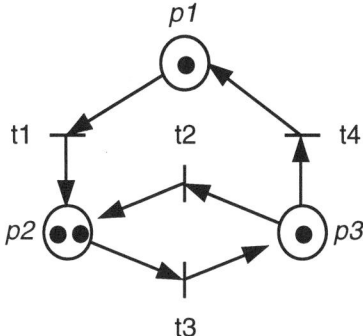

Figure 4.1.   Transitions 2 and 3 are uncontrollable.

Example.   The Petri net of Figure 4.1 contains two uncontrollable transitions: t_2 and t_3. Tokens in places p_2 and p_3 can not be prevented from freely traveling between these two places. However t_1 can be used to stop the introduction of new tokens into p_2 and p_3, and t_4 can be used to prevent tokens from leaving.

The constraint

$$\mu_3 \leq 1 \tag{4.1}$$

is inadmissible. The initial state of the plant $\mu_0 = \begin{bmatrix} 1 & 2 & 1 \end{bmatrix}^T$ satisfies the constraint, but the uncontrollable firing of t_3 would lead to the state $\mu = \begin{bmatrix} 1 & 1 & 2 \end{bmatrix}^T$, which violates (4.1). The constraint fails condition 2 of Definition 4.4.

The constraint

$$\mu_1 \leq 1 \tag{4.2}$$

is admissible. The current state of the plant satisfies the constraint, and for any state that satisfies the constraint, there is no firing of uncontrollable transitions that would lead to a state that does not satisfy it. The marking of p_1 is effected only by the firings of transitions t_1 and t_4, both of which are controllable.

If $L\mu_p \leq b$ is inadmissible, then it is desirable to find another constraint $L'\mu_p \leq b'$ such that $L'\mu_p \leq b'$ is an admissible constraint, and for all μ_p such that $L'\mu_p \leq b'$, $L\mu_p \leq b$ is also true.

Example.   In the example above, we could replace constraint (4.1) with

$$\mu_2 + \mu_3 \leq 1 \tag{4.3}$$

This constraint is admissible according to Definition 4.4, and all reachable states that satisfy (4.3) also satisfy (4.1). Thus constraint (4.1) could be enforced by designing a controller for constraint (4.3) using the technique of section 3.2. Unfortunately a

---

[2] Another way of stating that a constraint is admissible is to say that it is "control invariant."

controller designed this way may not be maximally permissive. The method of handling uncontrollable/unobservable transitions in chapter 5 follows along these lines, but it also includes the idea of finding *all constraints* $L'\mu_p \leq b'$ that meet the criteria above. A controller that enforces the disjunction of these inequalities will still be a relatively simple structure while allowing for a high degree of plant freedom.

Research into "vector discrete event systems" [Li and Wonham, 1993, Li and Wonham, 1994, Li and Wonham, 1995] is related to these issues in that they also seek to find ways to transform constraints into new forms that contain only admissible markings. Their work relevant to this issue is summarized in the following section.

4.4   Vector Discrete Event Systems

Important contributions to the area of uncontrollable transitions in discrete event system control have been made by [Li and Wonham, 1993, Li and Wonham, 1994, Li and Wonham, 1995]. Their primary contribution relating to this book involves a method for transforming a set of general constraints into a set that accounts for uncontrollable transitions in the plant. The transformation is accomplished by analytically solving an integer linear programming problem. This section contains a summary and discussion of this work.

Li and Wonham introduce and base their work around a DES model they call the vector discrete event system (VDES). The VDES is, at least in one of its forms, equivalent to a Petri net. The discussion here will refer only to the Petri net form of the VDES, which is the form used for most of the results on uncontrollability. Before proceeding with the main contribution, the following definitions are made. The definitions are illustrated using the Petri net of Figure 4.2.

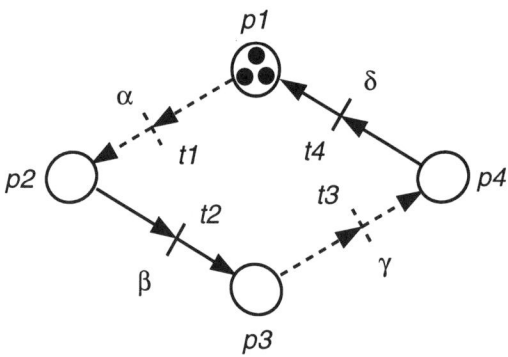

Figure 4.2.   A Petri net with uncontrollable transitions α, and γ.

G : The Petri net. A four place / four transition Petri net is shown in Figure 4.2.

$G_{uc} \in G$: The uncontrollable portion of the Petri net. In Figure 4.2, the uncontrollable transitions and associated arcs are shown as dashed rather than solid lines. G_{uc} is made up of p_1 and p_2 connected through transition t_1 and p_3 and p_4 connected through transition t_3.

Σ : The set of events. In Figure 4.2, $\Sigma = \{\alpha, \beta, \gamma, \delta\}$. One event is associated with the firing of each transition.

$\Sigma_{uc} \in \Sigma$: The set of uncontrollable events. $\Sigma_{uc} = \{\alpha, \gamma\}$ in Figure 4.2.

$L(G, \mu)$: The language of Petri net G starting at state μ. The language is composed of all possible sequences of events due to transition firings.

$w \in L(G, \mu)$: A valid or possible string of events in the PN G starting from the state μ.

D_{uc} : Uncontrollable columns of incidence matrix D. In Figure 4.2 we have:

$$D = \begin{bmatrix} -1 & 0 & 0 & 1 \\ 1 & -1 & 0 & 0 \\ 0 & 1 & -1 & 0 \\ 0 & 0 & 1 & -1 \end{bmatrix} \qquad D_{uc} = \begin{bmatrix} -1 & 0 \\ 1 & 0 \\ 0 & -1 \\ 0 & 1 \end{bmatrix} \qquad (4.4)$$

The control goal is to enforce a single linear on the marking behavior of G:

$$l^T \mu \leq b \qquad (4.5)$$

The method of Li and Wonham is used to transform (4.5) into a form that eliminates the possibility of uncontrollable transition firings leading to a violation of (4.5), i.e., transformation into an admissible constraint. Instead of using a Petri net structure, Li and Wonham define their controller as a boolean function that defines whether or not each controllable transition should be enabled given the current state.

A final state as a result of a sequence of transition firings can be represented as a single vector equation as follows. Given a sequence of N events, $w \in L(G, \mu)$, the final state is given by

$$\mu_N = \mu_0 + Dq_1 + Dq_2 + \ldots + Dq_N \quad = \quad \mu + D\underbrace{(q_1 + q_2 + \ldots q_N)}_{Q_w}$$

$$\mu_N \quad = \quad \mu + DQ_w \qquad (4.6)$$

Given (4.6), the control law $f^*(\alpha, \mu)$ is equal to 1 when event α should be enabled given the current state μ.

$$f^*(\alpha, \mu) = \begin{cases} 1 & \text{if } \mu + Dq_\alpha \in [P] \\ 0 & \text{else} \end{cases}$$

where

$$\begin{aligned} [P] \quad &= \quad \{\mu | (\forall w \in L(G_{uc}, \mu)), l^T(\mu + D_{uc}Q_{uc,w}) \leq b\} \\ &= \quad \{\mu | (l^T \mu + \max_{w \in L(G_{uc}, \mu)} l^T D_{uc}Q_{uc,w}) \leq b\} \end{aligned} \qquad (4.7)$$

In other words, the transition associated with the firing q_α is enabled if it is impossible that any subsequent firing of uncontrollable transitions could violate the constraint $l^T \mu \leq b$.

The maximization problem in (4.7) involves searching a nonconvex feasible region that may be difficult to solve. However if the uncontrollable portion of the plant net

G_{uc} contains no loops, then the set of possible strings $w \in L(G_i, \mu)$ can be simplified as follows.

$$\{Q_{uc,w} | w \in L(G_{uc}, \mu)\} = \{Q \in \mathbb{Z}^k | \mu + D_{uc}Q \geq 0, Q \geq 0\}$$

With this simplification of the feasible region, the set $[P]$ of allowed states becomes

$$[P] = \{\mu | l^T \mu + l D_{uc} Q^*(\mu) \leq b\} \tag{4.8}$$

where $Q^*(\mu)$ is the solution of

$$\max_Q l^T D_{uc} Q$$
$$\text{s.t.} \begin{cases} D_{uc}Q & \geq & -\mu \\ Q & \geq & 0 \text{ (integer)} \end{cases} \tag{4.9}$$

Symbolic solutions to linear integer programming problem (4.9) will yield Q^* as a function of μ.

The structure of the uncontrollable portion of the plant net, D_{uc}, will limit the types of functions that $Q^*(\mu)$ can be. In general, if the graph of the uncontrollable portion of the net has a tree structure, then $Q^*(\mu)$ will have a closed-form expression.

A Petri net is said to have a "general tree structure" if it is loop free and the places on one level of the Petri net graph act as sources for the transitions on the next level. These transitions then act as the exclusive source of tokens for the places in the next level of the graph. Places on one level receive tokens exclusively from the transitions one level back and deliver tokens to the transitions one level forward. The same level-to-level relationship holds for the transitions as well. Furthermore, transitions do not share input places, i.e., each place may contain no more than one output arc. Figure 4.3a shows an example of a general tree structure. A "tree structure of type 1" exists when the Petri net has a general tree structure, and every transition has at most one output arc. A "tree structure of type 2" exists when the Petri net has a general tree structure, and every transition has at most one input arc. Figs. 4.3b and 4.3c show examples of types 1 and 2 tree structures.

The following claims, proved in [Li and Wonham, 1994], are made based on the structure of D_{uc}, the uncontrollable portion of the plant Petri net:

1. If D_{uc} has a type 1 tree structure, then the set $[P]$ of allowed states has the form of a disjunction of finitely many linear constraints:

$$[P] = \bigvee_{i=1}^{t} c_i^T \mu_p \leq b \tag{4.10}$$

for some c_i and t.

2. If D_{uc} has a type 2 tree structure, then the set $[P]$ of allowed states has the form of a single linear constraint, in the same form as the original constraint $l^T \mu_p \leq b$:

$$[P] = c^T \mu \leq b \tag{4.11}$$

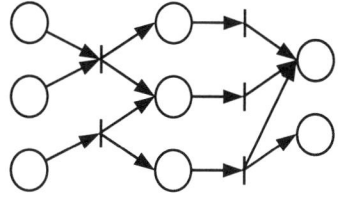

Figure 4.3a.   General tree structure.

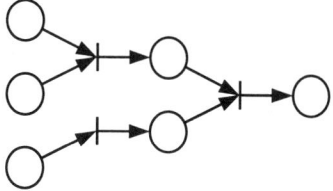

Figure 4.3b.   Type 1 tree structure.

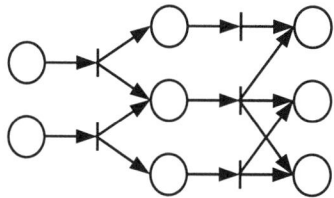

Figure 4.3c.   Type 2 tree structure.

for some c.

Clearly, when situation 2 exists, a maximally permissive controller that will insure that the state remains within $[P]$ can be efficiently implemented using the control method described in section 3.2. Similarly, situation 1 can be handled by controllers designed according to the procedure of section 5.4, which deals with disjunctions of linear constraints. However the "tree structure" of D_{uc} is not a necessary condition for $[P]$ to have either of the two forms described.

Example.   Given the Petri net of Figure 4.2 with uncontrollable transitions t_1 and t_3 and the incidence matrices D and D_{uc} given in (4.4), we will enforce the constraint

$$\mu_4 \le 1$$

so

$$l = [0\,0\,0\,1]^T \qquad\qquad b = 1$$

The allowable states are given by (4.8), which is found by solving (4.9):

$$\max \; [0\,0\,0\,1] \begin{bmatrix} -1 & 0 \\ 1 & 0 \\ 0 & -1 \\ 0 & 1 \end{bmatrix} \begin{bmatrix} q_1 \\ q_2 \end{bmatrix}$$

s.t.

$$\begin{bmatrix} -1 & 0 \\ 1 & 0 \\ 0 & -1 \\ 0 & 1 \end{bmatrix} \begin{bmatrix} q_1 \\ q_2 \end{bmatrix} \ge \begin{bmatrix} -\mu_1 \\ -\mu_2 \\ -\mu_3 \\ -\mu_4 \end{bmatrix} \qquad\qquad q_i \ge 0 \ (\text{integer})$$

Rewriting we have

$$\max \; q_2$$

s.t.

$$q_1 \geq -\mu_2 \tag{4.12}$$
$$q_2 \geq -\mu_4 \tag{4.13}$$
$$q_1 \leq \mu_1 \tag{4.14}$$
$$q_2 \leq \mu_3 \tag{4.15}$$

Constraints (4.12) and (4.13) are trivial because we already have $q_i, \mu_i \geq 0$. The set of allowable states will be defined by (4.14) and (4.15). Substituting into equation (4.8) we have

$$[P] = \{\mu|\mu_4 + l^T D_{uc} Q_u^* \leq 1\}$$
$$= \{\mu|\mu_4 + q_2^* \leq 1\}$$

where q_2^* is the solution of $\max_{q_2} q_2$ s.t. $q_2 \leq \mu_3$, i.e., $q_2^* = \mu_3$ and the set of allowable states is defined as

$$[P] = \{\mu|\mu_4 + \mu_3 \leq 1\} \tag{4.16}$$

The uncontrollable portion of this plant has a type 2 tree structure, and thus $[P]$ has the simple linear form that we would expect. The boolean control law is then

Control Law: $f^*(i, \mu) = \begin{cases} 1 & \mu + Dq_i \in [P] \\ 0 & \text{else} \end{cases} \quad 1 \leq i \leq 4$

A Petri net controller for this problem can be found by applying the synthesis technique of chapter 3 to the set of allowed states defined in (4.16). The controlled plant is shown in Figure 4.4.

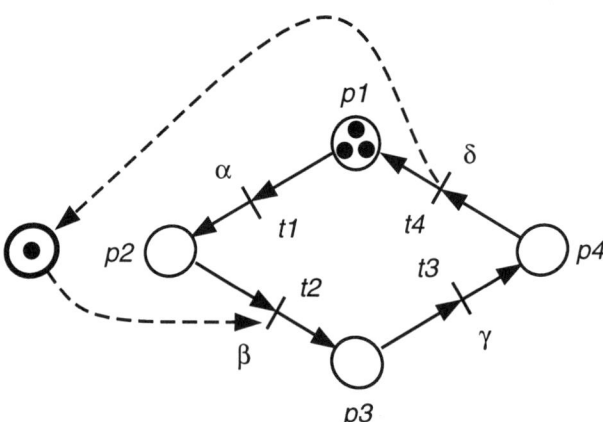

Figure 4.4.   The Petri net of Figure 4.2 with controller shown with dashed lines.

In order to obtain an analytic expression for $[P]$, it was necessary to solve the integer programming problem symbolically. If this is not possible or is too cumbersome, then it is necessary for the controller to numerically solve integer programs at every iteration of the VDES's evolution. This can be computationally expensive for large problems.

4.5   Petri Net Modeled Supervisors

Unlike the controllers of Li and Wonham, the supervisors used in this book are modeled by Petri nets, though this basic structure may be modified slightly as is done in section 5.4 for the enforcement of disjunctions of linear constraints. Uncontrollable and unobservable transitions can cause problems for PN based supervisors due to limitations in their modeling power, however Petri net supervisors are still useful for several reasons:

- Unified plant/controller models are elegant, facilitating implementation and closed loop system analysis.

- The evolution of Petri net models is inexpensive to compute, facilitating their use in real time control applications.

- Desirable Petri net qualities, such as the automatic handling of concurrent events, are maintained with unified plant/controller PN models.

- Though the decision power of a Petri net supervisor is not unlimited, a good variety of DES control problems can be effectively and efficiently solved through their use.

- Recognizing the controller as a Petri net facilitates understanding of what can and can not be done with the supervisor. This will become evident in the material below.

This section will be used to present theoretical results and analytical tools used in the synthesis of Petri net supervisors for plants with uncontrollable and unobservable transitions.

For an invariant based Petri net supervisor to be realizable on a plant with uncontrollable and unobservable transitions, the constraint it is enforcing must be admissible. The following proposition provides necessary and sufficient conditions for any state or event based constraint to be admissible. It involves analyzing the behavior of a maximally permissive controller constructed to supervise the given constraint. Here maximally permissive is used in the sense of chapter 3, where all transitions are assumed to be controllable. In this case, a maximally permissive controller only prevents firings that lead to states that directly violate the given constraint.

Proposition 4.5 General constraint admissibility. A constraint on the marking and/or firing behavior of a Petri net is *admissible* iff

1. The initial conditions of the plant satisfy the constraint, and

2. There exists a maximally permissive controller (constructed under the assumption that all transitions are controllable) that enforces the constraint and does not inhibit any uncontrollable transitions that would otherwise be enabled.

Proof. Clearly, if the initial conditions of a plant violate a constraint, then that constraint can not be enforced and is inadmissible according to condition 1 of Definition 4.4. Furthermore, if the constraint is admissible, then a maximally permissive controller would have no need to attempt to disable otherwise enabled uncontrollable transitions, as per Definition 4.4.

A maximally permissive controller will allow any reachable state or behavior that does not violate the constraint. Thus, if a maximally permissive controller never attempts to inhibit an otherwise enabled uncontrollable transition, then the constraint it is enforcing is admissible according to Definition 4.4.   ■

The following corollary relates Proposition 4.5 to the PN controllers of chapter 3 for linear place-constraints on a plant with incidence matrix D_p and initial marking μ_{p_0}.

Corollary 4.6 Place-constraint admissibility. A single vector constraint $l^T \mu_p \leq b$ is *admissible* iff the controller with incidence matrix $D_c = -l^T D_p$ and initial marking $\mu_{c0} = b - l^T \mu_{p0} \geq 0$ will never attempt to disable an uncontrollable transition that would otherwise be enabled.

Proof. The proof is by Proposition 4.5 since the given controller is constructed under the assumption that all transitions are controllable and is maximally permissive (see section 3.3).   ■

Corollary 4.6 deals with individual inequality constraints rather than the block form $L\mu_p \leq b$ because each of the inequalities in the block form can be handled independently. Certain constraints in $L\mu_p \leq b$ may be admissible, while others may not.

The following equations are from Theorem 3.2

$$D_c = -LD_p \tag{4.17}$$

$$\mu_{c_0} = b - L\mu_{p_0} \tag{4.18}$$

These equations show that it is possible to construct the incidence matrix D_c of a maximally permissive Petri net controller as a linear combination of the rows of the incidence matrix of the plant. Negative elements in D_c correspond to arcs from controller places to plant transitions. These arcs act to inhibit plant transitions when the corresponding controller places are empty, and thus they can only be applied to plant transitions that permit such external control. If, as in the previous sections, we group all of the columns of D_p that correspond to transitions that can not be controlled into the matrix D_{uc}, we obtain the following corollary.

Corollary 4.7 $l^T D_{uc} \leq 0$ implies admissibility. Given a plant with uncontrollable transitions described by the incidence matrix D_{uc} and a constraint $l^T \mu_p \leq b$, if

$$l^T D_{uc} \leq 0 \tag{4.19}$$

then the constraint is admissible for the given plant.

Proof. The proof follows from Corollary 4.6 and the construction of the Petri net controller whose incidence matrix is $D_c = -l^T D_p$ as described in section 4.1. Inequality (4.19) insures that the controller draws no arcs to uncontrollable transitions.

■

Example.   Corollary 4.7 can be used to verify the results from the example in section 4.3. Since transitions t_2 and t_3 are uncontrollable in the Petri net of Figure 4.1, D_{uc} is composed of the second and third columns of the plant incidence matrix.

$$D_p = \begin{bmatrix} -1 & 0 & 0 & 1 \\ 1 & 1 & -1 & 0 \\ 0 & -1 & 1 & -1 \end{bmatrix}$$
$$\underbrace{}_{D_{uc}}$$

Constraint (4.1) fails to meet condition (4.19) of the corollary.

$$\begin{bmatrix} 0 & 0 & 1 \end{bmatrix} D_{uc} = \begin{bmatrix} -1 & 1 \end{bmatrix}$$

Constraints (4.2) and (4.3) both meet condition (4.19) and are both admissible.

$$\begin{bmatrix} 1 & 0 & 0 \end{bmatrix} D_{uc} = \begin{bmatrix} 0 & 0 \end{bmatrix}$$
$$\begin{bmatrix} 0 & 1 & 1 \end{bmatrix} D_{uc} = \begin{bmatrix} 0 & 0 \end{bmatrix}$$

Remark. For most practical examples, constraints that fail condition (4.19) are inadmissible and will need to be transformed if they are to be enforced, however Corollary 4.7 provides only a sufficient condition for constraint admissibility. The following corollary indicates a situation where the controller may contain arcs to uncontrollable transitions in the plant, i.e., $l^T D_{uc} \nleq 0$, but the constraint is still admissible.

Corollary 4.8   Redundant place-constraints are admissible. Given the single vector constraint

$$l^T \mu_p \leq b$$

on the marking of a plant with incidence matrix D_p and initial marking μ_{p0}, if there exits a place invariant x such that for all reachable μ_p, $x^T \mu_p = x^T \mu_{p0}$ implies that $l^T \mu_p \leq b$ is also true, then the constraint is admissible.

Proof. The place invariant shows that $l^T \mu_p \leq b$ is always true due to the structure of the plant. A maximally permissive controller only disables an otherwise enabled transition when the firing of that transition would cause $l^T \mu > b$. But this situation can never occur given the natural evolution of the plant, so the controller will never disable an otherwise enabled transition, and the constraint is admissible according to Proposition 4.5.   ■

Remark. Corollary 4.8 refers to the case where all reachable markings of the plant already satisfy the given constraint (hence the use of the word "redundant," see also the redundant *GMEC's* of section 3.1). The practical use of this corollary is in conjunction with the material in section 7.2, which deals with constraints on a plant's firing vector. This technique is illustrated in the "three tank" example of section 8.6.2.

As discussed in section 4.2, it is illegal for the controller to change its state based on the firing of an unobservable transition, because there is no direct way for the controller to be told that such a transition has fired. Both input and output arcs from the controller places are used to change the controller state based on the firings of plant transitions. Let the matrix D_{uo} represent the incidence matrix of the unobservable portion of the Petri net. This matrix is composed of the columns of D_p that correspond to unobservable transitions, just as D_{uc} is composed of the uncontrollable columns of D_p.

Corollary 4.9 $l^T D_{uo} = 0$ implies admissibility. Given a plant with unobservable transitions described by the incidence matrix D_{uo} and a constraint $l^T \mu_p \leq b$, if

$$l^T D_{uo} = 0 \tag{4.20}$$

then the constraint is admissible.

Proof. As with Corollary 4.7, the proof follows from Corollary 4.6 and the construction of the Petri net controller whose incidence matrix is $D_c = -l^T D_p$ as described in section 4.2. Equation (4.20) insures that the controller draws no arcs to or from unobservable transitions. ∎

Remark. Corollaries 4.7 and 4.9 indicate that it is possible to observe a transition that we can not inhibit, but it is illegal to directly inhibit a transition that we can not observe.

Suppose, given a set of constraints $L\mu_p \leq b$, we construct the matrices LD_{uc} and LD_{uo} and observe that there are violations to conditions (4.19) and/or (4.20). Since the controller is made of a linear combination of the rows of D_p, it is interesting to consider the situation where we use the addition of further rows from D_{uc} in order to eliminate the positive elements of LD_{uc} and use rows from D_{uo} to eliminate the nonzero elements of LD_{uo}, i.e., if we are going to use a place invariant forming Petri net controller, what additions to the constraints would we need to make in order to eliminate positive elements from LD_{uc} and nonzero elements from LD_{uo}? What admissible constraints, of the form $L' \mu_p \leq b'$, will also maintain the original constraint $L\mu_p \leq b$? The following lemma appeared in [Moody et al., 1995b].

Lemma 4.10 Constraint transformation structure.

$$\text{Let } R_1 \in \mathbb{Z}^{n_c \times n} \text{ satisfy } R_1 \mu_p \geq 0 \,\forall\, \mu_p. \tag{4.21}$$
$$\text{Let } R_2 \in \mathbb{Z}^{n_c \times n_c} \text{ positive definite diagonal matrix} \tag{4.22}$$

If $L' \mu_p \leq b'$ where

$$
\begin{aligned}
L' &= R_1 + R_2 L \tag{4.23}\\
b' &= R_2(b+1) - 1 \tag{4.24}
\end{aligned}
$$

and $\mathbf{1}$ is an n_c dimensional vector of 1's, then $L\mu_p \leq b$.

Proof. The transformed constraint is $(R_1 + R_2 L)\mu_p \leq R_2(b+1) - 1$. Because all of the elements are integers, the inequality can be transformed into a strict inequality:

$$(R_1 + R_2 L)\mu_p < R_2(b + 1)$$

Because R_2 is diagonal and positive definite,

$$R_2^{-1} R_1 \mu_p + L\mu_p < b + 1$$

Assumptions (4.21) and (4.22) imply that all elements of the vector $R_2^{-1} R_1 \mu_p \geq 0$, therefore $L\mu_p \leq b$.   ∎

Lemma 4.10 shows a class of constraints, $L'\mu_p \leq b'$, which, if enforced, will imply that $L\mu_p \leq b$ is also enforced. The following lemma is used to show the conditions under which a particular set of constraints can be enforced on a particular Petri net.

Lemma 4.11 Initial condition check for transformed constraints. The constraint $L'\mu_p \leq b'$, where $L' \neq 0$ and b' are defined by (4.23) and (4.24), can be enforced on a Petri net with initial marking μ_{p_0} iff

$$0 \leq R_1 \mu_{p_0} \leq R_2(b + 1 - L\mu_{p_0}) - 1 \tag{4.25}$$

Proof. Substituting L' and b' into (4.25) gives $0 \leq b' - L'\mu_{p_0}$, which is equivalent to the condition $L'\mu_{p_0} \leq b'$ that states that the initial conditions of the plant must fall within the acceptable region of the constraints. Clearly, if a controller does exist, then the initial conditions of the plant must not violate the constraints. Furthermore, as shown in chapter 3 (see also [Moody et al., 1994, Yamalidou et al., 1996, Giua et al., 1992]), if the initial conditions lie within the acceptable region of the plant (inequality (3.9)), a controller to enforce the conditions can be computed with incidence matrix $D_c = -L'D_p$ and initial marking $\mu_{c_0} = b' - L'\mu_{p_0}$.   ∎

Theorem 4.12 combines Corollaries 4.9 and 4.7 with the conditions for creating a valid set of transformed constraints in Lemmas 4.10 and 4.11 to show how to construct a Petri net controller.

Theorem 4.12 Constraint transformation and supervisor synthesis. Let a plant Petri net with incidence matrix D_p be given with a set of uncontrollable transitions described by D_{uc} and a set of unobservable transitions described by D_{uo}. A set of linear constraints on the net marking, $L\mu_p \leq b$, are to be imposed. Assume R_1 and R_2 meet (4.21) and (4.22) with $R_1 + R_2 L \neq 0$ and let

$$\begin{bmatrix} R_1 & R_2 \end{bmatrix} \begin{bmatrix} D_{uc} & D_{uo} & -D_{uo} & \mu_{p_0} \\ LD_{uc} & LD_{uo} & -LD_{uo} & L\mu_{p_0} - b - 1 \end{bmatrix} \leq \begin{bmatrix} 0 & 0 & 0 & -1 \end{bmatrix} \tag{4.26}$$

Then the controller

$$D_c = -(R_1 + R_2 L)D_p = -L'D_p \tag{4.27}$$

$$\mu_{c_o} = R_2(b+1) - 1 - (R_1 + R_2 L)\mu_{p_o} = b' - L'\mu_{p_o} \tag{4.28}$$

exists and causes all subsequent markings of the closed loop system (3.6) to satisfy the constraint $L\mu_p \leq b$ without attempting to inhibit uncontrollable transitions and without detecting unobservable transitions.

Proof. According to (4.17) and (4.18), equations (4.27) and (4.28) define a controller that enforces the constraint $L'\mu_p \leq b'$. Lemma 4.10 shows that if assumptions (4.21) and (4.22) are met then a controller that enforces a particular constraint $L'\mu_p \leq b'$ will also enforce the constraint $L\mu_p \leq b$. The fourth column of inequality (4.26) indicates that the condition in lemma 4.11 is satisfied, thus the controller exists and the control law can be enforced. The first column of (4.26) indicates that $L'D_{uc} \leq 0$, thus condition (4.19) is satisfied and no controller arcs are drawn to the uncontrollable transitions. The second and third columns of (4.26) indicate that $L'D_{uo} = 0$, thus condition (4.20) is satisfied and no arcs are drawn between the controller places and the unobservable plant transitions.   ∎

Note that $\begin{bmatrix} R_1 & R_2 \end{bmatrix}$, which is used to describe the constraint transformation, multiplies from the left in (4.26), thus these matrices represent the use of rows from D_{uc} to eliminate positive elements from LD_{uc}, and the use of rows from D_{uo} to zero the elements of LD_{uo}, as discussed above. An illustration of some of these techniques used for handling uncontrollable transitions can be found in the "cat and mouse" problem of section 8.1.2 and the "unreliable machine" example of section 8.3. The piston rod robotic assembly cell of section 8.4 also includes unobservable transitions.

Remark. The procedure above allows for the complete transformation of one set of linear constraints, $L\mu_p \leq b$, into another, $L'\mu_p \leq b'$. However the techniques described in section 4.4 indicate that when it is possible to do a simple linear transformation of the constraints, it is not necessary to change the right-hand-side of the constraint inequality, i.e., $b' = b$. The following example is used to illustrate why $b' = R_2(b+1) - 1$ in the development above and that it is sometimes necessary to change both the left-hand and the right-hand sides of the constraint inequality in order to obtain a linear form for the optimal transformed constraint.

Example.   Consider the Petri net of Figure 4.5. The control goal is to impose the condition

$$\mu_4 \leq 2 \tag{4.29}$$

or, in matrix form,

$$\underbrace{\begin{bmatrix} 0 & 0 & 0 & 1 \end{bmatrix}}_{L} \mu_p \leq \underbrace{2}_{b}$$

The incidence matrices of the plant and the uncontrollable portion of the plant are

$$D_p = \begin{bmatrix} -1 & 0 & 0 & 1 \\ 0 & -1 & 0 & 2 \\ 2 & 1 & -2 & 0 \\ 0 & 0 & 1 & -2 \end{bmatrix} \qquad D_{uc} = \begin{bmatrix} 0 \\ 0 \\ -2 \\ 1 \end{bmatrix}$$

Transition t_3 is uncontrollable, thus D_{uc} is the third column of D_p.

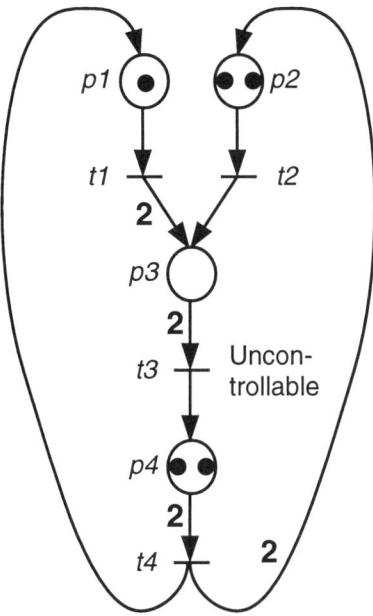

Figure 4.5.   Should transition t_2 be allowed to fire if the constraint is $\mu_4 \leq 2$?

Checking the quantity LD_{uc} we see that $LD_{uc} = 1$ is positive, thus a direct enforcement of (4.29) would involve an attempt to control the uncontrollable transition. A simple row operation is then performed on $\begin{bmatrix} D_{uc} \\ LD_{uc} \end{bmatrix}$ to find R_1 and R_2 that satisfy (4.21), (4.22), and (4.26):

$$R_1 = \begin{bmatrix} 0 & 0 & 1 & 0 \end{bmatrix}$$
$$R_2 = 2$$

which yields the following transformation of the constraint:

$$\left(\begin{bmatrix} 0 & 0 & 1 & 0 \end{bmatrix} + 2 \begin{bmatrix} 0 & 0 & 0 & 1 \end{bmatrix} \right) \mu_p \leq 2(2+1) - 1$$

or

$$\mu_3 + 2\mu_4 \leq 5 \tag{4.30}$$

Now we will obtain a similar transformation using the method described in section 4.4. The single uncontrollable transition contains one input and one output arc, giving it a type 2 tree structure. The optimal constraint transformation should then be a linear function of μ_p given by (4.8) that requires us to solve the integer linear programming problem (4.9):

$$\max_{Q} LD_{uc}Q = \max Q$$

$$\text{s.t.} \quad \begin{cases} \begin{bmatrix} \mu_1 \\ \mu_2 \\ \mu_3 \\ \mu_4 \end{bmatrix} + \begin{bmatrix} 0 \\ 0 \\ -2 \\ 1 \end{bmatrix} Q \geq 0 \\ \\ \quad Q \geq 0 \text{ (integer)} \end{cases}$$

The only nontrivial constraint above is $\mu_3 - 2Q \geq 0$. Thus Q^*, the maximum possible value of Q, is given by

$$Q^* = \text{floor } \frac{1}{2}\mu_3$$

which, according to (4.8), gives the constraint

$$\text{floor } \frac{1}{2}\mu_3 + \mu_4 \leq 2 \tag{4.31}$$

which is indeed the optimal control law for this problem, but it is not expressed as a linear integer-valued inequality. If we naively multiplied both sides of (4.32) by 2 in order to eliminate the fraction, we would have

$$\mu_3 + 2\mu_4 \leq 4 \tag{4.32}$$

which is a linear constraint, but is not maximally permissive. Consider the plant in the state shown in Figure 4.5. If we are to insure that μ_4 remains less than or equal to 2, should the firing of transition t_2 be allowed? Clearly the answer is yes, and (4.30) and (4.31) would both allow this, but condition (4.32) would not, thus it is not maximally permissive.

Because all of the elements in the inequalities are integers, constraint (4.30) is equivalent to the nonlinear constraint (4.31). That is, the integer valued feasible regions of (4.30) and (4.31) are identical, but (4.30) is in a form which can be instantly translated into a Petri net controller, while (4.31) can not. This is the reason why the procedure outlined above includes a transformation for b to b'.

Remark. It is possible that uncontrollable transitions in a plant might make a particular constraint impossible to realize. In this case it may still be possible to find R_1 and R_2 such that they meet the assumptions, however the transformed constraint $L'\mu_p \leq b'$ will be trivial. For example, consider the Petri net in Figure 4.6. Suppose that we wish to limit the number of tokens that enter p_2, i.e., the untransformed constraint is $\mu_2 \leq b$. If the single transition is uncontrollable, then we will obtain a transformed constraint of $\mu_1 + \mu_2 \leq b$, which is already the case if there are b or fewer tokens in the net and is impossible if the net starts with more than b tokens.

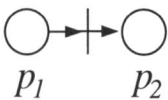

$$p_1 \qquad p_2$$

Figure 4.6.   A Petri net that yields a trivial constraint transformation.

5 CONSTRAINT TRANSFORMATION AND CONTROLLER SYNTHESIS

The mathematical tools and propositions of the previous chapter provide the backbone for the control synthesis procedures presented here. In section 5.1, two techniques are introduced for generating transformations of linear constraints that will meet the conditions of Theorem 4.12. Methods for characterizing all admissible constraints for plants with uncontrollable and unobservable transitions are developed in section 5.2. These techniques are then employed in sections 5.3 and 5.4 for the synthesis of controllers that enforce the disjunctions of linear inequalities allowing for a high degree of plant freedom. The technique for enforcing nonconvex constraints in section 5.4 requires a modification of the controller net's transition enabling rule, however this modification does not decrease the computational efficiency of the closed loop system model, and it leaves many of the analytical results and tools intact.

5.1 Computing Constraint Transformations

The usefulness of Theorem 4.12 for specifying controllers to handle plants with uncontrollable and unobservable transitions lies in the ease in which the matrices R_1 and R_2 can be generated. Section 5.1.1 shows how the information in the proposition can be converted into an integer linear programming problem (ILP), and section 5.1.2 proposes a method based on matrix row operations.

5.1.1   An Integer Linear Program for calculating R_1 and R_2

It is possible to convert the conditions in Theorem 4.12 into an integer linear programming problem (ILP) in the standard form of

$$\min_{x} x^T c$$
$$\text{s.t.} \begin{cases} x^T A &= d \\ x &\geq 0 \text{ (integer)} \end{cases} \tag{5.1}$$

We will consider only a single constraint on the system; multiple constraints can be handled individually and independently. Thus $n_c = 1$, L and R_1 are vectors, and b and R_2 are scalars.

To satisfy condition (4.21), we can specify that $R_1 \geq 0$, since we know that $\mu_p \geq 0$. In fact, it is necessary to specify that all of the elements of R_1 are greater than or equal to zero if the markings of the plant places are unbounded or their bounds are not known.

Condition (4.22) states that we need $R_2 > 0$. To obtain variables that fit the conditions for x in (5.1), define

$$R_2' = R_2 - 1 \tag{5.2}$$

Since R_2 is an integer, $R_2' \geq 0$ implies $R_2 > 0$.

Substituting the new variable R_2' into condition (4.19) yields

$$R_1 D_{uc} + R_2' L D_{uc} \leq -L D_{uc}$$

A vector of slack variables, $R_3 \geq 0$, is introduced to convert the inequality into an equality.

$$R_1 D_{uc} + R_2' L D_{uc} + R_3 = -L D_{uc} \tag{5.3}$$

R_3 is a row vector like R_1 but with dimension equal to the number of columns of D_{uc} (n_{uc}, the number of uncontrollable transitions).

Substituting R_2' into condition (4.20) gives

$$R_1 D_{uo} + R_2' L D_{uo} = -L D_{uo} \tag{5.4}$$

which is already in the form of an equality, so an additional slack variable is unnecessary.

R_2' is now substituted into the condition given in Lemma 4.11 that indicates whether the given constraint transformation can be implemented.

$$R_1 \mu_{po} + R_2'(L\mu_{po} - (b+1)) \leq b - L\mu_{po} \tag{5.5}$$

Let

$$R = \begin{bmatrix} R_1 & R_2' & R_3 \end{bmatrix}$$

and the ILP can now be defined as

$$\min_{R} \left(z(R) = R \begin{bmatrix} \mu_{p_0} \\ L\mu_{p_0} - b - 1 \\ 0 \end{bmatrix} \right)$$

$$\text{s.t.} \begin{cases} R \begin{bmatrix} D_{\text{uc}} & D_{\text{uo}} \\ LD_{\text{uc}} & LD_{\text{uo}} \\ I & 0 \end{bmatrix} = -L \begin{bmatrix} D_{\text{uc}} & D_{\text{uo}} \end{bmatrix} \\ R \geq 0 \text{ (integer)} \end{cases} \quad (5.6)$$

which is in the form of (5.1).

After solving (5.6), if the minimum of the objective function $z^* = z(R^*)$ is greater than $b - L\mu_{p_0}$ then the problem can not be solved as there are no values of R_1 and R_2 that will satisfy the condition in Lemma 4.11. If the minimum is less than or equal to $b - L\mu_{p_0}$, then transform R_2' back into R_2 and generate the controller using the formulae in Theorem 4.12.

Remark. There may be minor difficulties encountered when using this method of generating R_1 and R_2. For a controller to exist, we need the objective function of the ILP, $z(R) \leq b - L\mu_{p_0}$, however it not clear that we should attempt to *minimize* this function. Oftentimes, this function will have an unbounded minimum, making it necessary for the designer to introduce an additional constraint in order to achieve a bounded solution. Care must also be taken such that the ILP does not yield the pathological transformation $L' = R_1 + R_2L = 0$, when there are other nonzero possibilities for L'.

5.1.2   Matrix Row Operations

The Algorithm

It is possible to obtain appropriate constraint transformations by performing row operations on a matrix containing the uncontrollable and unobservable columns of the plant incidence matrix. The computational part of this procedure involves little more than the integer triangularization of a matrix, and thus it is simpler to compute R_1 and R_2 using this method than by using the ILP presented in the previous section. Before presenting the algorithm itself, the following terms are clarified:

D_{uc} : An $n \times n_{\text{uc}}$ matrix consisting of the columns of the plant incidence matrix D_p that correspond to transitions that are uncontrollable, *but which may be observed*. n_{uc} is the number of these transitions.

D_{uo} : An $n \times n_{\text{uo}}$ matrix consisting of the columns of D_p that are unobservable (just as defined in previous sections).

In the discussions above, D_{uc} may have included columns that were unobservable as well as uncontrollable, but here all of the columns of D_{uc} are observable. Conditions (4.19) and (4.20) show that unobservability implies stricter demands than uncontrollability, and in fact, any transition labeled as unobservable is also uncontrollable. D_{uc} is defined here as being strictly observable so that we can relax our requirements when dealing with this matrix. Algorithm 5.1 presents the procedure for determining R_1 and R_2.

As was done in section 5.1.1, we shall insure that condition (4.21) is met by making $R_1 \geq 0$. In terms of row operations, this means that elements in rows are eliminated

Algorithm 5.1 (Constraint Transformation)
Input: $L \in \mathbb{Z}^{n_c \times n}, b \in \mathbb{Z}^{n_c}, D_{uc} \in \mathbb{Z}^{n \times n_{uc}}, D_{uo} \in \mathbb{Z}^{n \times n_{uo}}, \mu_{p_o} \in \mathbb{Z}^n$
if $(LD_{uc} \leq 0$ and $LD_{uo} = 0)$ then
     $R_1 := 0_{n_c \times n}, R_2 := I_{n_c \times n_c}$
else
$$M := \begin{bmatrix} D_{uc} & D_{uo} & \\ LD_{uc} & LD_{uo} & I \end{bmatrix}$$
     Let $M(i,j)$ represent the $(i,j)^{\text{th}}$ element of matrix M.
     Zero all positive elements in the LD_{uc} portion of
       M following the procedure in Algorithm 5.2.
     if $M(n+1 \ldots n + n_c, 1 \ldots n_{uc})$ has any positive elements
       then
          FAILURE: Controller arcs were introduced by the
             row operations into the uncontrollable portion
             of the plant, and they can not be eliminated
             due to lack of negative elements in D_{uc}.
     end if
     Zero all elements in the LD_{uo} portion of the M
       matrix following the procedure in Algorithm 5.3.
     if $M(n+1 \ldots n + n_c, n_{uc} + 1 \ldots n_{uc} + n_{uo})$ has any nonzero
       elements then
          FAILURE: The unobservable portion of the plant
             contains arcs which can not be eliminated.
     end if
     $R_1 := M(n+1 \ldots n+n_c, n_{uc} + n_{uo} + 1 \ldots n_{uc} + n_{uo} + n)$
     $R_2 := M(n+1 \ldots n+n_c, n_{uc} + n_{uo} + n + 1 \ldots n_{uc} + n_{uo} + n + n_c)$
end if
$L' := R_1 + R_2 L$
$b' := R_2(b+1) - 1$
if $L'\mu_{p_o} > b'$ then
     FAILURE: Control law is infeasible.
end if
Output: L' and b'.

strictly through addition, never through subtraction, and that rows can be multiplied only by positive integers. The procedure for zeroing out the elements in a column of numbers that have the opposite sign of the "pivot" is given in Algorithm 5.4.

Algorithm 5.1 insures that condition (4.22) is met because the procedure for choosing the "pivot" elements never picks from the LD_{uc} and LD_{uo} portions of the M matrix. Combined with the zeroing procedure of Algorithm 5.4, these steps insure that the R_2 portion of the M matrix is diagonal with strictly positive elements.

Algorithm 5.2 (Elimination of positive elements from D_{uc})

```
Input:   Working matrix M ∈ ℤ^((n+nc)×(nuc+nuo+n+nc))
```

$i := 1$

while $i \leq \min(n_{uc}, n)$

   if $M(i...n, i)$ has any negative elements then

      Let j be the index of a row in $M(i...n, i)$ which
         contains a negative element.

      Exchange rows i and j of M

      Use the negative pivot value at $M(i,i)$ to
         eliminate all positive integers in the column
         $M(i...n + n_c, i)$ (See Algorithm 5.4.)

   else if $M(n + 1...n + n_c, i)$ has any positive elements
   then

      FAILURE: A controller arc can not be eliminated
         because there are no negative elements in the
         corresponding column of D_{uc}.

   end if

   $i := i + 1$

end while

Output M and i

Algorithms 5.2 and 5.3 (called by Algorithm 5.1) are used to make sure that the transformed constraints meet conditions (4.19) and (4.20). The feasibility check at the end of Algorithm 5.1 directly tests the condition of Lemma 4.11 to insure that the controller does exist.

The instructions for picking positive or negative elements to act as pivots in the two main loops are left specifically vague. Different methods of choosing the pivot will lead to different constraint transformations. *It would be possible, for instance, to find a basis for all valid constraint transformations by repeating the procedures in Algorithm 5.1 whenever there was a choice of more than one pivot for a given column.*

Generality of the Algorithm

Even if all the possible pivots are used in Algorithm 5.2, there may still be other R_1 and R_2 values that yield transformed constraints that meet condition (4.19). Algorithm 5.2 forces all positive elements in the LD_{uc} portion of the matrix to go to zero. However condition (4.19) states that we need $LD_{uc} \leq 0$, so the question is, is it ever desirable or necessary to transform positive elements of LD_{uc} into negative elements? Should Algorithm 5.2 incorporate this ability? At this time, this question can not be answered definitely, however the following points, starting with the lemma below, indicate that a procedure that causes positive elements of LD_{uc} to become negative is undesirable.

Lemma 5.1 A single controller place can possess either an input or an output arc (or neither) to any given plant transition, but it will never contain both, i.e., a place invariant based controller contains no self-loops.

Algorithm 5.3 (Zeroing of all elements in D_{uo})

```
Input:  Working matrix M ∈ ℤ^((n+n_c)×(n_uc+n_uo+n+n_c)) and i
while i ≤ min(n_uc + n_uo, n)
  if M(i...n, i) contains any negative elements then

      Let j be the index of a row in M(i...n, i) which
        contains a negative element.
      Exchange rows i and j of M
      k := 1
  else
      k := 0
  end if
  if M(i...n, i) has any positive elements then
      Let j be the index of a row in M(i...n, i) which
        contains a positive element.
      Exchange rows i + k and j of M
      Use the positive pivot value at M(i    +    k, i) to
        eliminate all
        negative integers in the column M(i + k...n + n_c, i)
        (See Algorithm 5.4.)
  end if
  if k = 1 then
      Use the negative pivot value at M(i, i) to
        eliminate all positive integers in the column
        M(i...n + n_c, i)
  end if
  if M(n + 1...n + n_c, i) has nonzero elements then
      FAILURE: Controller arc(s) can not be eliminated
        from the unobservable portion of the plant.
  end if
  i := i + 1
end while
Output M
```

Proof. The proof is simply by construction of the controller incidence matrix. The incidence matrix of the invariant forming controller for establishing the constraint $L\mu_p \le b$ is defined by $D_c = -LD_p$. The matrix D_c^+ consists of the positive elements in D_c, D_c^- consists of the negative, and all other elements are zero, i.e., there are never any cancellations between D_c^+ and D_c^- used in forming D_c. See [Yamalidou et al., 1996] as well. ∎

A positive element in the matrix LD_{uc} means that the controller would draw an arc to a transition in the uncontrollable portion of the plant. Lemma 5.1 tells us

---

Algorithm 5.4 (Column Zeroing)
```
Input:   Working matrix M ∈ Z^(n+nc)×(nuc+nuo+n+nc) and pivot
   position (p, j).
i := p + 1
while i ≤ n + nc
   if M(i, j) is opposite M(p, j) in sign then
      while M(i, j) ≠ 0
         if |M(p, j)| > |M(i, j)| then
            d := floor (−M(p, j)/M(i, j))
            if (mod(M(p, j), M(i, j)) = 0) then
               d := d − 1 (Don't zero-out the pivot!)
            end if
            M(p, .) := M(p, .) + dM(i, .) (row operation)
         else
            d := floor (−M(i, j)/M(p, j))
            M(i, .) := M(i, .) + dM(p, .) (row operation)
         end if
      end while
   end if
   i := i + 1
end while
Output M
```

---

that this element is not the result of battling input and output arcs. The controller wishes to output to this transition, and it has no reason to receive input from this transition. Algorithm 5.2 would then be used to eliminate the controller's interaction with the transition altogether, if it were uncontrollable. But what would happen if row operations were used to change this output transition of the controller into an input transition?

Place invariant forming controllers use their places to take up the slack that exists in the inequality $l^T \mu_p \leq b$. Specifically, the marking of the controller, μ_c, is given by

$$\mu_c = b - l^T \mu_p$$

Negative elements in D_c correspond to transitions that decrease the slack μ_c and lead to the controller placing inhibitions on the plant's behavior. Positive elements correspond to transitions that increase the slack and give the plant a greater margin of freedom. A positive element in $l^T D_{uc}$ means that the firing of the corresponding transition will decrease the slack. If this positive number is changed, via row operations and the transformation of l to l', to a negative number, it means that an event that previously indicated a decrease in the slack and the freedom of the plant has become one that indicates an increase in slack and plant freedom. The fact that the quantity $b - l^T \mu_p$ has decreased while the quantity $b' - l'^T \mu_p$ has increased seems to indicate that the controller in the transformed system is concerning itself with matters that have gone beyond the original request of insuring that $l^T \mu_p \leq b$. This is the reason that Algorithm 5.2 makes no attempt to find row operations that would lead to this situation, however the ideas presented here are not a proof that it is never necessary to convert output arcs into input arcs in order to handle uncontrollable transitions.

5.2   Structure of Admissible Constraints and Controls

Given a plant with uncontrollable/unobservable transitions, it is useful to seek methods for transforming inadmissible constraints into admissible ones, but it is also logical to ask, in general, what are the admissible constraints for this plant? Is there a way to characterize all or most of these constraints? Section 5.2.1 provides a method for just such a characterization. Sections 5.2.2 and 5.3 show how this characterization can be used to synthesize controllers.

5.2.1   Characterization of Admissible Constraints

As in the previous sections, let the matrix D_{uo} represent the incidence matrix of the unobservable portion of the Petri net. It is illegal for the controller $D_c = -LD_p$ to contain any arcs in the unobservable portion of the net, thus an admissible set of constraints will satisfy

$$LD_{uo} = 0 \tag{5.7}$$

as indicated in Corollary 4.9.

Any L that satisfies (5.7) will lie within the kernel of D_{uo}. Let X satisfy

$$XD_{uo} = 0 \tag{5.8}$$

where X is an integer matrix with dimension $(n - \text{rank } D_{uo}) \times n$. The rows of X form a linearly independent basis for the kernel of D_{uo} (X is full rank). The process of finding X is equivalent to finding the place invariants (an algorithm appears in [Martinez and Silva, 1980]) of the unobservable portion of the plant Petri net. All admissible constraints must lie within the basis described by the rows of X, and thus can be formed as linear combinations of these rows, i.e., every admissible constraint can be described by $k^T X$ where k is an integer vector with dimension $(n - \text{rank } D_{uo})$. In general, the coefficient matrix of any set of feasible constraints $L' \in \mathbb{Z}^{n_c \times n}$ can be written

$$L' = KX \tag{5.9}$$

where $K \in \mathbb{Z}^{n_c \times (n - \text{rank } D_{uo})}$. Equation (3.10) can then be used to calculate the incidence matrix of the controllers that will enforce these constraints:

$$D_c = -KXD_p \tag{5.10}$$

A characterization of all admissible constraints and controls is not quite as transparent for the case of uncontrollable transitions as it is for unobservable ones. For unobservable transitions we have an equality, $LD_{uo} = 0$, that must be satisfied, but for uncontrollable transitions it is an inequality, $LD_{uc} \leq 0$, so we can not simply find the kernel of D_{uc}. In this case, the following equality can be formed

$$LD_{uc} + \Delta = 0$$

where Δ is a positive semidefinite diagonal matrix of slack variables. The previous equation is rewritten

$$[\, L \quad \Delta \,] \begin{bmatrix} D_{uc} \\ I \end{bmatrix} = 0$$

A kernel X, solving

$$X \begin{bmatrix} D_{uc} \\ I \end{bmatrix} = 0$$

can then be used to construct a basis for all admissible linear constraints that may be placed on the plant. X must be computed so as to insure that the elements of X that correspond to Δ are nonnegative. Because the matrix whose kernel is being found contains I, the elements corresponding to Δ in X will be independent of each other: any negative numbers in this portion of X can be corrected by a row-multiplication of -1. After insuring that none of the slack variables are negative, all admissible constraint matrices L can be found in the linear combinations of X that leave nonnegative values in the slack columns. An algorithm for performing this procedure appears in section 5.3.3, but first, the simpler case of constraint transformation for plants with unobservable transitions is discussed below.

5.2.2   Constraint Transformations for Unobservability

Suppose we have a set of constraints $L\mu \leq b$ such that $LD_{uo} \neq 0$. It is necessary to create new constraint matrices (L', b') with two properties.

1. $L'D_{uo} = 0$

2. $\forall \mu_p, L'\mu_p \leq b' \to L\mu_p \leq b$

Property 1 is necessary to insure that the new controller will not utilize the unobservable transitions. The characterization of all such matrices L' was described above. Property 2 indicates that the new constraints must be at least as restrictive as the original ones. Lemma 4.10 from section 4.5 is used to deal with this condition.

To perform the transformation, it is necessary to determine values for the matrices R_1 and R_2 defined in lemma 4.10 that meet assumptions (4.21) and (4.22). Computational techniques for determining these matrices were given in section 5.1, however it is possible for a designer to determine the values of R_1 and R_2 by using the kernel of D_{uo}. Combining equations (5.9) and (4.23) we see that

$$L' = KX = R_1 + R_2 L$$

The designer should multiply each constraint in L by some positive integer (which will determine the diagonal elements in R_2) and add new positive coefficients (which will determine R_1) such that the new constraint is a linear combination of the rows of X. This process will yield the L' matrix, and b' can be calculated using R_2 and equation (4.24). An illustration of this procedure appears in section 8.4.3 in the example of the piston rod robotic assembly cell.

5.3   Admissible Constraints and Controller Synthesis

5.3.1   Illustrations of Issues

The place invariant controller method yields maximally permissive supervisors for enforcing linear constraints of the form $L\mu_p \leq b$. When these constraints are transformed, because of the uncontrollability and unobservability of certain transitions, into $L'\mu_p \leq b'$, the invariant based control method will still yield a maximally permissive realization of the transformed constraint. Unfortunately, the new constraint itself may not represent the most permissive admissible control law corresponding to the original constraint. The maximally permissive admissible constraint associated with a linear predicate on the plant's marking may be a nonlinear predicate that can not be optimally controlled by a standard invariant based controller. However, the modified controller structure of section 5.4 may work as it permits the enforcement of nonconvex disjunctions of constraints.

At this time, a complete description of the conditions under which an optimal transformation of linear constraints is another set of linear constraints is unknown. [Li and Wonham, 1994] have shown that when the uncontrollable portion of the plant has a "type 1 tree structure," the optimal transformation will be a disjunction of linear predicates, while a type 2 structure will yield a linear transformation (see section 4.4). However these conditions are only sufficient, not necessary.

Three simple examples are presented below to illustrate the kinds of situations that can occur in deriving a proper constraint transformation.

Example 1. Unique L', Unique Controller, Optimal Control

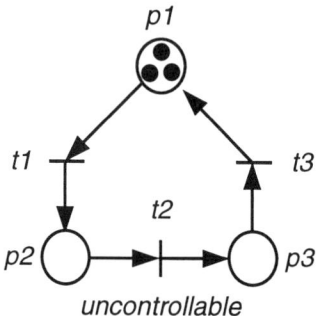

Figure 5.1.   A Petri net that yields an optimal constraint transformation.

The incidence matrix of the Petri net in Figure 5.1 is

$$D_p = \begin{bmatrix} -1 & 0 & 1 \\ 1 & -1 & 0 \\ 0 & 1 & -1 \end{bmatrix}$$

Transition t_2 is uncontrollable and the uncontrollable portion of the net has a type 2 tree structure. The control goal is

$$\underbrace{\begin{bmatrix} 0 & 0 & 1 \end{bmatrix}}_{L} \mu_p \le b$$

Algorithm 5.1 indicates that all valid transformations of L to L', ignoring transformations that would change output arcs to input arcs as discussed at the end of section 5.1, are given by

$$L' = k \begin{bmatrix} 0 & 1 & 1 \end{bmatrix}$$

where $k \in \mathbb{Z}, k > 0$. The multiplication of the constraint inequality by a constant does not change the set of allowable states, so we shall set $k = 1$. Thus, for this example, there is a unique transformation of L to L'. The incidence matrix of the corresponding controller is, of course, also unique.

$$D_c = \begin{bmatrix} -1 & 0 & 1 \end{bmatrix}$$

and this is the optimal controller, and L' is the optimal constraint transformation, as we should expect since the uncontrollable portion of the plant has a type 2 tree structure.

Example 2.   Infinite L', Unique Controller, Optimal Control

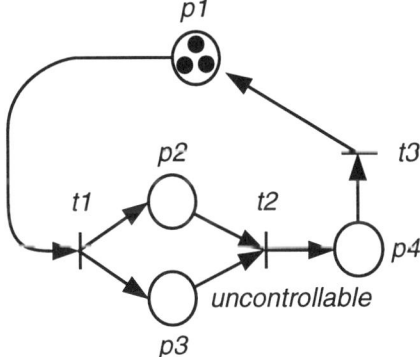

Figure 5.2.   A Petri net that yields many constraint transformations but a unique, optimal controller.

The incidence matrix of the Petri net in Figure 5.2 is

$$D_p = \begin{bmatrix} -1 & 0 & 1 \\ 1 & -1 & 0 \\ 1 & -1 & 0 \\ 0 & 1 & -1 \end{bmatrix}$$

Transition t_2 is uncontrollable and the uncontrollable portion of the net does not have a type 2 tree structure (it is type 1). The control goal is

$$\underbrace{\begin{bmatrix} 0 & 0 & 0 & 1 \end{bmatrix}}_{L} \mu_p \le b$$

An infinite number of different L' values can be obtained by Algorithm 5.1

$$L' = \begin{bmatrix} 0 & k_1 & k_2 & k_1 + k_2 \end{bmatrix}$$

where the constants $k_1, k_2 \in \mathbb{Z}, k_1, k_2 \ge 0$, and at least one constant is nonzero. However, when we obtain the controller for these transformed constraints we get

$$D_c = k \begin{bmatrix} -1 & 0 & 1 \end{bmatrix}$$

where $k = k_1 + k_2$. Multiplying the incidence matrix of the controller does nothing to change the actual control law, it only changes the number of tokens the controller uses in its internal representation of slack. Thus the controller for this problem is unique, and in fact, the control law it enforces is optimal (maximally permissive).

Example 3. Infinite L', Infinite Controllers, Suboptimal Control

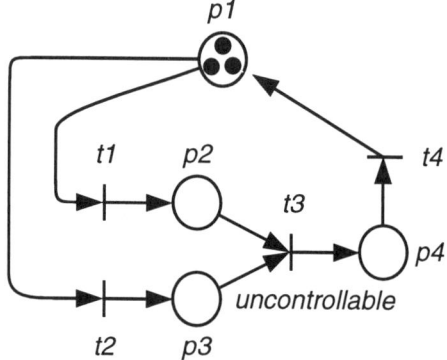

Figure 5.3.   A Petri net that yields many constraint transformations and standard invariant based controllers that all lead to suboptimal control.

The incidence matrix of the Petri net in Figure 5.3 is

$$D_p = \begin{bmatrix} -1 & -1 & 0 & 1 \\ 1 & 0 & -1 & 0 \\ 0 & 1 & -1 & 0 \\ 0 & 0 & 1 & -1 \end{bmatrix}$$

Transition t_3 is uncontrollable and the uncontrollable portion of the net has the same tree structure as in the previous example. The control goal is

$$\underbrace{\begin{bmatrix} 0 & 0 & 0 & 1 \end{bmatrix}}_{L} \mu_p \le b$$

As in Example 2, an infinite number of different L' values can be obtained by Algorithm 5.1

$$L' = \begin{bmatrix} 0 & k_1 & k_2 & k_1 + k_2 \end{bmatrix}$$

where $k_1, k_2 \in \mathbb{Z}, k_1, k_2 \geq 0$, and at least one constant is nonzero. This time, each constraint yields a different controller

$$D_c = \begin{bmatrix} -k_1 & -k_2 & 0 & k_1 + k_2 \end{bmatrix}$$

All of these controllers result in the enforcement of suboptimal control laws. The optimal control law is

$$\mu_4 + \min(\mu_2, \mu_3) \leq b \tag{5.11}$$

which can not be realized by a standard linear place invariant forming controller.

Constraint (5.11) is equivalent to the following disjunction of two linear constraints.

$$(\mu_2 + \mu_4 \leq b) \vee (\mu_3 + \mu_4 \leq b) \tag{5.12}$$

This constraint can be enforced using the procedure for disjunctions of linear constraints presented in section 5.4. The associated controller would contain, as usual, two slack places, one for each of the inequalities in the disjunction. However, instead of insisting that both control places always remain nonnegative, the condition is relaxed such that, at any time, only one control place must be nonnegative. This means that the supervisor is no longer an ordinary Petri net, but the state evolution laws remain the same, and the transition enabling condition is only slightly modified. The details of these modified controllers are covered in section 5.4; the remainder of 5.3 explains how nonconvex constraints like (5.12) can be generated automatically.

5.3.2   Generation of Nonconvex Control Laws

Section 5.2.1 provides a means of characterizing all *admissible* constraints of the form $l^T \mu_p \leq b$ that may be enforced on a given plant with uncontrollable transitions. This characterization is based, in part, on computing the kernel of the matrix $\begin{bmatrix} D_{uc} \\ I \end{bmatrix}$.

Lemma 4.10 provides a means of characterizing linear inequalities $l'^T \mu_p \leq b'$ such that

$$\forall \mu_p, l'^T \mu_p \leq b' \rightarrow l^T \mu_p \leq b$$

where l' is a linear transformation of l

$$l' = R_1 + R_2 l$$

and R_1 and R_2 meet assumptions (4.21) and (4.22).

Given an inadmissible constraint $l^T \mu_p \leq b$, a possibly *nonconvex control law*[1] for the enforcement of this constraint can be synthesized using the algorithm presented in the following subsection. The algorithm follows this basic procedure:

---

[1] A control law is called nonconvex when the set of states reachable under the control law is a nonconvex set. $L\mu_p \leq b$ always describes a convex set of states, while disjunctions of inequalities are, in general, nonconvex.

1. Find all inequalities $l'^T \mu_p \leq b'$ that are

   (a) Valid transformations of $l^T \mu \leq b$ according to Lemma 4.10 and

   (b) Admissible constraints according to the theory developed in section 5.2.1.

   There may be an infinite number of inequalities that meet these two requirements, but they may be expressed with a finite number of inequalities that form a linearly independent basis.

2. Construct the controller incidence matrices associated with these constraints using $D_c = -l'^T D_p$. There may be fewer unique D_c rows than there are inequalities as was seen in Example 2 of section 5.3.1.

3. Enforce the *logical union* of the individual control actions by constructing controllers according to the procedure of section 5.4.

The procedure above is similar to the idea of the *supremal controllable sublanguage* [Wonham and Ramadge, 1987, Ramadge and Wonham, 1989] from the supervisory control literature. In both cases, all of the valid behaviors of the plant are characterized based on the plant's structure and the desired, constrained behavior, and the supervisor is then used to insure that the behavior of the plant is limited to this set of admissible behaviors.

To say that the procedure above will always result in a *maximally* permissive control law, the following two points would have to be proved.

1. The maximally permissive control law associated with a plant and constraint $l^T \mu_p \leq b$ can always be expressed as the disjunction of other linear state inequalities.

2. The transformation procedure in Lemma 4.10 covers all valid constraint transformations, i.e., if for all $\mu_p \geq 0$ such that $l'^T \mu_p \leq b'$, $l\mu_p \leq b$ is also true, then (l', b') can be expressed as a linear function of (l, b) according to the rules and assumptions of Lemma 4.10.

[Li and Wonham, 1994] have shown that condition 1 is true when the uncontrollable portions of a plant have a certain "tree structure" (see section 4.4). But for the general case, the answer is not known.

5.3.3   Supervisor Synthesis Algorithm

An algorithm is presented below to implement the procedure described in section 5.3.2. To combine the set of valid constraint transformations from Lemma 4.10 with the set of admissible constraints described in section 5.2.1 we introduce the transformation parameters, R_1 and R_2, into the admissibility condition, $l^T D_{uc} \leq 0$,

$$(R_1 + R_2 l)^T D_{uc} \quad \leq \quad 0$$

where the elements of the vector R_1 are nonnegative and the scalar $R_2 > 0$ to insure compliance with assumptions (4.21) and (4.22) of Lemma 4.10. A slack variable is now introduced.

$$(R_1 + R_2 l)^T D_{uc} + \Delta = 0$$

$$\begin{bmatrix} R_1^T & R_2 & \Delta \end{bmatrix} \begin{bmatrix} D_{uc} \\ l^T D_{uc} \\ I \end{bmatrix} = 0$$

The procedure for finding valid transformations will involve finding the kernel of

$$\begin{bmatrix} D_{uc} \\ l^T D_{uc} \\ I \end{bmatrix}$$

and manipulating the result such that it yields nonnegative R_1 and Δ values and positive R_2 values.

The steps of the algorithm are illustrated by the plant from Example 3 of section 5.3.1. The incidence matrix and constraint for this problem are repeated here.

$$D_p = \begin{bmatrix} -1 & -1 & 0 & 1 \\ 1 & 0 & -1 & 0 \\ 0 & 1 & -1 & 0 \\ 0 & 0 & 1 & -1 \end{bmatrix} \qquad \underbrace{\begin{bmatrix} 0 & 0 & 0 & 1 \end{bmatrix}}_{l^T} \mu_p \le b$$

Constructing $l^T D_p$

$$l^T D_p = \begin{bmatrix} 0 & 0 & 1 & -1 \end{bmatrix} \tag{5.13}$$

shows that there is a positive number in the third column, which is the column associated with the uncontrollable transition, and the constraint is inadmissible. The algorithm for constructing a controller follows.

1. *Find the kernel, M, of*

$$\begin{bmatrix} D_{uc} \\ l^T D_{uc} \\ I \end{bmatrix}$$

   and insure that all elements of M that correspond to the slack variable Δ are nonnegative.

   For the example, D_{uc} is the third column of D_p.

$$\text{kernel} \quad \begin{bmatrix} 0 \\ -1 \\ -1 \\ 1 \\ 1 \\ 1 \end{bmatrix} = M = \left[\begin{array}{cccc|c|c} 1 & 0 & 0 & 0 & 0 & 0 \\ 0 & -1 & 1 & 0 & 0 & 0 \\ 0 & 1 & 0 & 1 & 0 & 0 \\ 0 & 1 & 0 & 0 & 1 & 0 \\ 0 & 1 & 0 & 0 & 0 & 1 \\ \end{array} \right]$$
$$\underbrace{\phantom{\begin{array}{cccc}1 & 0 & 0 & 0\end{array}}}_{R_1^T} \; \underbrace{}_{R_2} \; \underbrace{}_{\Delta}$$

Each row of M represents a single kernel vector. The first four columns correspond to R_1^T, the fifth column corresponds to R_2, and the last column corresponds to the slack variable Δ. All of the slack elements are nonnegative. Multiplication of a row by -1 would have been used to change the sign of any negative slack values.

2. *Check if M contains any rows where $R_1 = 0$ and $R_2 = 1$. In this case the original constraint $l^T \mu_p \leq b$ is admissible. Construct a supervisor for the constraint using the standard procedure and STOP.*

The constraint in the example was already determined to be inadmissible, and none of the rows of M meets the conditions of this step in the algorithm.

3. *Eliminate rows from M that contain independent R_1 values, i.e., rows where $R_2 = 0$ and R_1 contains a single 1 with all other elements equal to 0.*

Rows that meet this condition represent places that may be independently added to the control law and are unnecessary to the search for valid transformations. Rows 1 and 5 of the example M meet this condition and are eliminated.

$$
M = \begin{bmatrix}
0 & -1 & 1 & 0 & 0 & 0 \\
0 & 1 & 0 & 1 & 0 & 0 \\
0 & 1 & 0 & 0 & 1 & 0
\end{bmatrix}
$$

4. *Perform row operations on M to insure that all R_2 values are positive, while maintaining nonnegative Δ values. If this is not possible, then the algorithm has failed, STOP.*

Rows 1 and 2 of M contain zeros in the R_2 column, so row 3 is added to each of these to make them positive.

$$
M = \begin{bmatrix}
0 & 0 & 1 & 0 & 1 & 0 \\
0 & 2 & 0 & 1 & 1 & 0 \\
0 & 1 & 0 & 0 & 1 & 0
\end{bmatrix}
$$

5. *Eliminate any remaining negative numbers in the matrix through further row operations. If a row contains negative numbers that can not be eliminated, then remove the entire row from M. If M is reduced to an empty matrix then the algorithm has failed, STOP.*

All negative numbers have been removed from the M matrix of the example.

6. *Create L' using the rows of M. Each row of M starts with R_1^T and R_2, each row of L' is constructed as $(R_1 + R_2 l)^T$. Construct b' as $R_2(b + 1) - 1$ in the same manner.*

$$
L' = \begin{bmatrix}
0 & 0 & 1 & 1 \\
0 & 2 & 0 & 2 \\
0 & 1 & 0 & 1
\end{bmatrix}
$$

7. *Compute $L'D_p$ and remove rows that are multiples of other rows.*

$$L'D_p = \begin{bmatrix} 0 & 1 & 0 & -1 \\ 2 & 0 & 0 & -2 \\ 1 & 0 & 0 & -1 \end{bmatrix} \quad \begin{array}{l} \text{Row 2 and row 3} \\ \text{are multiples} \Rightarrow \end{array} \begin{bmatrix} 0 & 1 & 0 & -1 \\ 1 & 0 & 0 & -1 \end{bmatrix}$$

8. *Eliminate any row in $L'D_p$ that contains a negative number in the same position that $l^T D_p$ contained a positive number. Do not leave $L'D_p$ as an empty matrix due to this step.*

   This step is included because of the intuition that if the original constraint were admissible and yielded a supervisor that would cause the slack-token count to diminish upon the firing of a certain transition, then the supervisor for the modified constraint should not experience an increase in the slack-token count upon the firing of the same transition (see the discussion at the end of section 5.1.2).

   In the example, $L'D_p$ contains negative numbers in the fourth column, but the fourth column of $l^T D_p$ is also negative, so no rows are deleted.

9. *Construct a supervisor Petri net*

$$D_c = -L'D_p \qquad \mu_{c_0} = b' - L'\mu_{p_0}$$

   If D_c contains more than one row, then its enabling rules should be treated as in section 5.4 to make it enforce a disjunction of linear inequalities rather than a conjunction.

$$D_c = \begin{bmatrix} 0 & -1 & 0 & 1 \\ -1 & 0 & 0 & 1 \end{bmatrix}$$

   D_c contains two rows, and should be implemented using the rules of section 5.4. Note that, for this example, the controller is maximally permissive.

5.4   Enforcing Disjunctions of Linear Constraints

The inequality

$$L\mu_p \leq b$$

represents the logical intersection, or conjunction, of n_c separate linear inequalities. The feasible solutions to the inequalities form a convex region [Fang and Puthenpura, 1993], and the behavior of a Petri net can be restricted to this region by adding further PN structures to the net as was shown above. A logical union, or disjunction, of linear constraints is, in general, nonconvex and can not be enforced with maximal permissivity on a Petri net through the use of other Petri net structures due to the linear nature of reachable PN state spaces. This claim is proved in the following lemma and proposition.

Lemma 5.2 A consequence of the PN enabling condition. Given a Petri net with initial marking μ_0 and two firing vectors q_1 and q_2, if the transition firings represented by q_1 or q_2 may occur, from the initial conditions, at least two times consecutively, i.e.,

$$D^-(kq_1) \leq \mu_0 \tag{5.14}$$

and

$$D^-(kq_2) \leq \mu_0 \qquad (5.15)$$

for $1 \leq k \leq 2$, then q_1 and q_2 may both be fired, from the initial conditions, one after the other or concurrently.

Proof. If $D^-(q_1 + q_2) \leq \mu_0$, then q_1 and q_2 are both feasible firing vectors and may be fired either consecutively or concurrently. Since we are dealing with strictly nonnegative values,

$$D^-(q_1 + q_2) \leq \max(2D^-q_1, 2D^-q_2)$$

and according to assumptions (5.14) and (5.15)

$$\max(2D^-q_1, 2D^-q_2) \leq \mu_0$$

Thus two valid firings in a Petri net, each of which may occur at least twice, implies that both firings may occur one after the other or even simultaneously.   ∎

Proposition 5.3 A Petri net controller can not always enforce the disjunction of two linear inequalities on the PN state with maximal permissivity.

Proof. The proposition is proved simply by providing a common example for which no PN controller will be maximally permissive. Given a plant with marking μ_p and a controller with marking μ_c, a closed loop system is formed with marking $\mu^T = [\mu_p^T \ \mu_c^T]$. The controller may be any Petri net connected in any way to the plant Petri net. Given initial conditions $\mu_0^T = [\mu_{p0}^T \ \mu_{c0}^T]$ and a closed loop incidence matrix D, let

$$\begin{bmatrix} \mu_{p1}(k) \\ \mu_{c1}(k) \end{bmatrix} = \begin{bmatrix} \mu_{p0} \\ \mu_{c0} \end{bmatrix} + D(kq_1) \geq 0 \qquad (5.16)$$

and

$$\begin{bmatrix} \mu_{p2}(k) \\ \mu_{c2}(k) \end{bmatrix} = \begin{bmatrix} \mu_{p0} \\ \mu_{c0} \end{bmatrix} + D(kq_2) \geq 0 \qquad (5.17)$$

be reachable states, for $1 \leq k \leq 2$, such that enabling conditions (5.14) and (5.15) are satisfied.

The controller must enforce the constraint

$$(l_1^T \mu_p \leq b_1) \vee (l_2^T \mu_p \leq b_2) \qquad (5.18)$$

Let

$$l_1^T \mu_{p1}(k) \leq b_1, \qquad l_2^T \mu_{p1}(k) > b_2$$
$$l_1^T \mu_{p2}(k) > b_1, \qquad l_2^T \mu_{p2}(k) \leq b_2$$

for $1 \leq k \leq 2$, and let

$$\mu_{p3} = \frac{1}{2}(\mu_{p1}(2) + \mu_{p2}(2)) \geq 0$$

be an integer-valued vector such that

$$l_1^T \mu_{p3} > b_1, \qquad l_2^T \mu_{p3} > b_2$$

A maximally permissive controller will permit transition firings that lead to states $\mu_{p1}(k)$ or $\mu_{p2}(k)$. Lemma 5.2 indicates that the state obtained through the firing of both q_1 and q_2 is reachable. The value of this state is calculated:

$$
\begin{aligned}
\begin{bmatrix} \mu_{p0} \\ \mu_{c0} \end{bmatrix} + Dq_1 + Dq_2 &= \frac{1}{2} \left(\begin{bmatrix} \mu_{p0} \\ \mu_{c0} \end{bmatrix} + 2Dq_1 \right) + \frac{1}{2} \left(\begin{bmatrix} \mu_{p0} \\ \mu_{c0} \end{bmatrix} + 2Dq_2 \right) \\
&= \frac{1}{2} \begin{bmatrix} \mu_{p1}(2) \\ \mu_{c1}(2) \end{bmatrix} + \frac{1}{2} \begin{bmatrix} \mu_{p2}(2) \\ \mu_{c2}(2) \end{bmatrix} \\
&= \begin{bmatrix} \mu_{p3} \\ \frac{1}{2}(\mu_{c1}(2) + \mu_{c2}(2)) \end{bmatrix} \geq 0
\end{aligned}
\tag{5.19}
$$

Thus the forbidden plant state μ_{p3} is reachable and a maximally permissive Petri net controller for constraint (5.18) does not exist.   ■

The maximally permissive enforcement of a nonconvex constraint requires that the supervisor is equipped to handle, model, and incorporate the nonconvexity within its own evolution rules. Proposition 5.3 showed that nonconvex constraints can not, in general, be enforced with maximal permissivity on an ordinary Petri net using an external ordinary Petri net as a supervisor. This section will show how a slight modification to the evolution rules of the controller net can be made such that it will act as a maximally permissive supervisor for a class of nonconvex constraints.

The simple rules that govern ordinary Petri net behavior are what help to make the PN model so attractive both for analysis and implementation. The reluctance to modify this model for the enforcement of nonconvex constraints on PN plants is overcome for the following reasons.

1. The ability to handle the disjunction of linear constraints as well as their conjunction is a powerful advancement in the utility of the method and is necessary for the proper solution of problems in many applications.

2. Disjunctions of linear constraints are important for the enforcement of linear constraints under conditions in which transitions may be uncontrollable or unobservable. This issue is very important in DES control and is covered in section 5.3.

3. The modified rules for controller state evolution that enable the controller to enforce disjunctions as well as conjunctions of linear inequalities on the plant state space involve only a slight modification of the ordinary transition enabling rule. Analysis and implementation are very similar to that of ordinary Petri nets.

The following disjunction of linear inequalities is to be enforced on the marking of a plant with initial marking $\mu_{p_0} \in \mathbb{Z}^n$ (all elements nonnegative) and incidence matrix $D_p \in \mathbb{Z}^{n \times m}$.

$$\bigvee_{i=1}^{n_c} l_i^T \mu_p \leq b_i \tag{5.20}$$

where $l_i \in \mathbb{Z}^n$ and $b_i \in \mathbb{Z}$. Let

$$D_{c_i} = -l_i^T D_p \qquad (5.21)$$

and

$$\mu_{c_{io}} = b_i - l_i^T \mu_{p_o} \qquad (5.22)$$

for $1 \le i \le n_c$. This procedure is the same as detailed in section 3.2, thus each pair (D_{c_i}, μ_{c_i}) is a maximally permissive PN supervisor for enforcing the constraint $l_i^T \mu_p \le b_i$. However if all of these supervisor elements were to be simultaneously enforced on the plant, the reachable plant states would lie in the conjunction of the inequalities in (5.20), rather than their disjunction.

To enforce a disjunction of constraints, at least one of the inequalities $l_i^T \mu_p \le b_i$ must be true at every transition firing iteration of the net's evolution. Let

$$L = \begin{bmatrix} l_1 & l_2 & \cdots & l_{n_c} \end{bmatrix}^T$$

so that

$$D_c = -LD_p \qquad (5.23)$$

and

$$\mu_{c_o} = b - L\mu_{p_o} \qquad (5.24)$$

which is identical to the controller construction from section 3.2. However, the enabling rule for the controller portion of the net must be changed such that it insures that at least one of the inequalities is being obeyed at all times rather than all of them at all times.

A firing vector q is valid (indicates the firing of an enabled transition) iff

$$D_p^- q \le \mu_p \qquad (5.25)$$

and

$$\mu_{c_i} + D_{c_i} q \ge 0 \text{ for some } i \in 1 \ldots n_c \qquad (5.26)$$

Inequality (5.25) is the standard PN enabling condition for a plant that may include transitions with self loops. The enabling condition for the controller (5.26) does not include any $D_{c_i}^-$ terms because controllers constructed according to the rules in section 3.2 do not contain self loops. Condition (5.26) may also be written

$$\max_{i=1}^{n_c} (\mu_{c_i} + D_{c_i} q) \ge 0 \qquad (5.27)$$

Note that it is still true that

$$L\mu_p + \mu_c = b \qquad (5.28)$$

However, unlike the standard nonnegative slack variables from before, many of the elements of this μ_c may be negative. The restriction placed by condition (5.27) insures that at least one of the elements is nonnegative, and thus at any time, at least one of the inequalities in (5.20) is being obeyed; allowed states lie in the union of all the feasible regions of the individual linear inequalities.

Proposition 5.4 Maximal permissivity of disjunction-enforcing controllers. A controller constructed according to (5.23) and (5.24) using enabling rule (5.27) is a maximally permissive supervisor for the enforcement of constraint (5.20) on the plant (D_p, μ_{c_0}) if

$$\mu_{c_{i_0}} \geq 0 \text{ for some } i \in 1 \ldots n_c. \tag{5.29}$$

Proof. If condition (5.29) is not met, then the initial conditions of the plant violate the constraint (5.20) according to equation (5.24).

Equation (5.28) shows that the state space of the closed loop system being outside the bounds of constraint (5.20) is equivalent to the situation when all the elements of μ_c are negative. However this is the only condition that is prevented by enabling rule (5.27). The only time the controller will intervene to disable a transition is when the firing of that transition would cause a direct violation of constraint (5.20), and thus the supervisor is maximally permissive.

■

Remark. In terms of Proposition 5.3, a disjunction-enforcing controller could allow some negative numbers in the elements of μ_{c1} and μ_{c2} in equations (5.16) and (5.17). Thus the vector $\mu_{c1}(2) + \mu_{c2}(2)$ in equation (5.19) could contain all negative numbers making the forbidden state μ_{p3} unreachable.

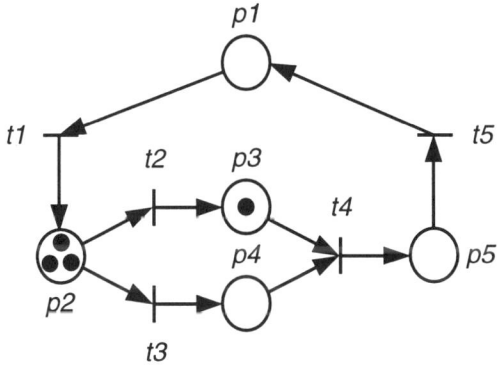

Figure 5.4.   The Petri Net plant for the example of section 5.4.

Example.   The following constraint is to be imposed on the behavior of the net in Figure 5.4.

$$(\mu_3 \leq 1 \text{ OR } \mu_4 \leq 1) \text{ AND } \mu_1 \leq 2 \tag{5.30}$$

The first two inequalities in (5.30) are involved in a disjunction. A controller place with associated arcs is constructed for each of the two inequalities. The rules for the construction of these two places is identical to the procedure of section 3.2, however the enabling rule for transitions connected to these places will obey (5.27) rather than the standard PN firing rules.

The final inequality in (5.30) will also have a PN controller created for it according to the standard rules. However, the relationship of this inequality to the previous two inequalities is a conjunction rather than a disjunction, so the transitions connected to it will follow the standard enabling rules.

The supervised system is shown in Figure 5.5. Places c_1 and c_2 are grouped together to indicate that they are combined to implement a single disjunction. The marking of either of these places may become negative, but at all times at least one of them must be nonnegative to insure that the disjunction of the first two inequalities in (5.30) is being obeyed. The final controller place, c_3, behaves like a standard PN place: it's marking must always be nonnegative. The combined action of the supervisor is a maximally permissive enforcement of (5.30).

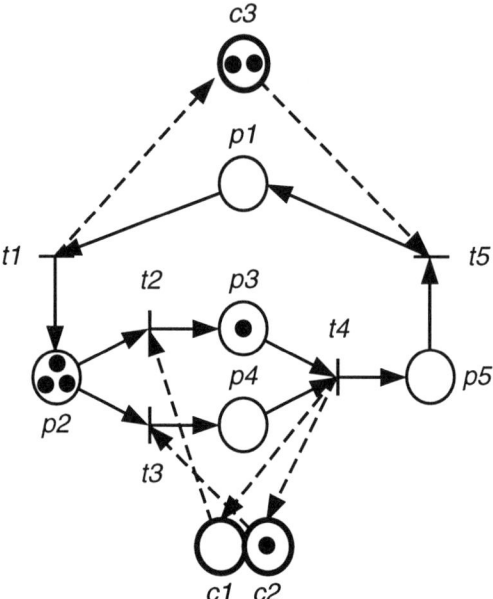

Figure 5.5.   The PN of Figure 5.4 with a supervisor for enforcing constraint (5.30).

6 RESOURCE MANAGEMENT AND DEADLOCK AVOIDANCE

The management and safe allocation of finite resources is one of the primary concerns for the supervision of discrete event systems. Section 6.1 shows how plant models can be easily and automatically modified to include these finite resource constraints using invariant based controllers.

The mismanagement of shared resources is one of the most common causes of deadlock for many systems. When a DES is modeled by a Petri net, the net is said to be *deadlocked* if no transition in the net is able to fire. A net is called *live* if every transition can be, eventually, fired again and again. A deadlock-free net may not necessarily be completely live. The supervision procedure of chapter 3 allows the control designer to restrict the marking behavior of a Petri net in a way that is maximally permissive. This means that the constraint inequalities are obeyed literally; if the feasible region of these inequalities includes a state in which no transition is enabled, the supervisor will allow the Petri net to enter this deadlock state.

The detection and avoidance of deadlock as well as the verification and insurance of liveness are important areas of research in net theory as well as DES control. [Corbett and Avrunin, 1995] have examined this issue and presented some solutions for the problem when the DES is modeled as an automaton. There are several contributions to solving or dealing with liveness and deadlock issues with Petri nets that fit very well into the supervision method of this book. Details of these approaches are presented later in this section.

Important structural and dynamic conditions relating to deadlock and liveness are described in section 6.2.1. This section includes a procedure developed by [Lautenbach and Ridder, 1994] for using transition invariants to determine the liveness of a class of nets. Several researchers have studied deadlock avoidance in their work on the modeling and control of manufacturing systems; these include [Ezpeleta et al., 1995, Bogdan and Lewis, 1997, Lewis et al., 1995, Huang et al., 1995, Tacconi et al., 1996, Huang et al., 1996, Xing et al., 1996]. Many of these results are relevant to invariant based control since they reduce to the enforcement of linear inequalities on the plant state. This is discussed in the first part of section 6.3. A less application-specific approach from [Barkaoui and Abdallah, 1995], that works seamlessly with invariant based controllers, is presented in the second part, which includes a look at the work of [Sreenivas, 1997a, Sreenivas, 1997b] regarding the existence of liveness-enforcing supervisors.

6.1   Modeling of Finite Resources

A finite resource is a tool or material with limited supply that is required by one or more agents for the completion of a job or to carry out some action. The availability of finite resources places implicit constraints on feasible actions within a system. These constraints can be written as linear inequalities on the state. Let b_i be the total number of available units for resource i. Let R_i be a set of places associated with finite resource i. Every token in the places making up the set R_i represents the use of one of the resources. A linear constraint on the marking can then be written

$$\sum_{\mu_j \in R_i} \mu_j \leq b_i \tag{6.1}$$

and can enforced by an invariant based supervisor.

Suppose that a resource suddenly becomes available or the number of available resources changes in some other way during the operation of the plant. This situation could be handled by modifying the token count in the appropriate controller slack place, i.e., the number b_i on the right hand side of (6.1) could be modified dynamically. Dynamic modifications to b in a constraint inequality will not change the structure of the controller. The arcs and their associated weights will remain the same. The only change would be in the marking of a controller place to correspond to the new slack value. Though this scheme would work, it is not very elegant from the point of view of Petri nets, i.e., tokens should not appear and disappear from a net without the corresponding firing of transitions.

Figure 6.1 shows how the resource controller places can be augmented with two uncontrollable transitions and a place in order to model the loss of finite resources while maintaining the standard Petri net framework of the model. Under normal operation, the token in the resource place is used to permit the firing of the plant transition. However the loss of the finite resource now corresponds to the firing of an uncontrollable transitions which robs the "resource is available" place of its token and stalls the operation of the plant. Another uncontrollable transition is then used to replace the missing token when the resource once again becomes available.

Finite resource constraints are illustrated in the example of the piston rod robotic assembly cell of section 8.4.

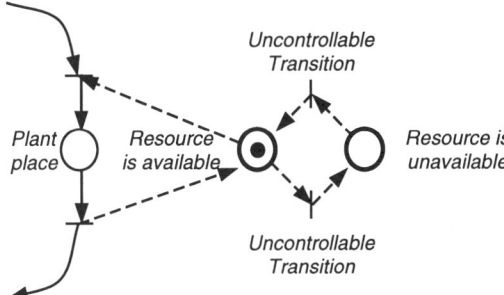

Figure 6.1.   Modeling the loss of a finite resource using uncontrollable transitions.

6.2   Conditions for Liveness

6.2.1   Controlled Siphons and Commoner's Theorem

Siphons (see section 2.3) are of particular interest in the area of deadlock avoidance; once a siphon becomes emptied of tokens, it will forever remain empty and all of the transitions that receive input arcs from these places will be dead. [Barkaoui and Abdallah, 1995] have introduced the notion of a controlled siphon.

Definition 6.1   For a Petri net with initial marking μ_0, a **controlled siphon** is a siphon that remains marked for all markings reachable from μ_0.   ■

A controlled siphon may be either trap-controlled or invariant-controlled. A *trap-controlled siphon* contains a trap that is initially marked, thus preventing the siphon from ever losing all of its tokens. An *invariant-controlled siphon's* marking is guaranteed by the presence of a place invariant. If the constant weighted sum of markings indicated by a place invariant insures that a siphon will never lose all of its tokens, then that siphon is invariant-controlled. Given an invariant vector x, with elements $x_1 \ldots x_n$, and a siphon S in a plant with initial marking μ_{p_0}, the invariant controls the siphon iff

1. $x_i > 0 \rightarrow p_i \in S$

2. $x_i < 0 \rightarrow p_i \notin S$

3. $x^T \mu_{p_0} > 0$

Example.   The Petri net of Figure 6.2 contains two siphons, $S_1 = \{p_2, p_3\}$ and $S_2 = \{p_2, p_3, p_4\}$, where S_1 is the only minimal siphon since $S_1 \subset S_2$. The net contains no traps, so clearly the minimal siphon is not trap-controlled, however an analysis of the net's behavior reveals that this siphon will never be emptied.

The net contains a single place invariant:

$$\mu_2 + \mu_3 - \mu_4 = 1$$

where μ_i is the marking of place p_i. Thus $\mu_2 + \mu_3 \geq 1$ is always true and the siphon is invariant-controlled.

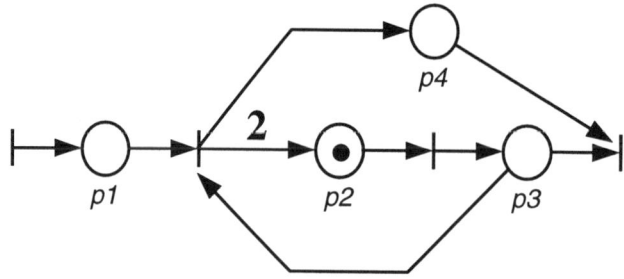

Figure 6.2.   A Petri net with a controlled siphon but no trap.

The following propositions relate controlled siphons to deadlock freedom and liveness.

Proposition 6.2 Deadlock condition. A deadlocked Petri net contains at least one empty siphon. Any net with an unmarked siphon is not live.

Proof. See [Desel and Esparza, 1995, Reisig, 1985, Murata, 1989].   ∎

Proposition 6.3 Deadlock-freedom. A Petri net is *deadlock-free* if every siphon in the net is a controlled siphon.

Proof. The standard form for this proposition [Desel and Esparza, 1995, Reisig, 1985, Murata, 1989] deals only with marked traps within every siphon. It is easily extended to the more general idea that every siphon is controlled (by a trap or an invariant).   ∎

Proposition 6.4 Commoner's Theorem. An extended free choice (EFC) Petri net is *live* if and only if every siphon in the net is trap-controlled.

Proof. See [Desel and Esparza, 1995, Reisig, 1985, Murata, 1989].   ∎

Remark. Note that (extended) free choice nets do not require the notion of invariant controlled siphons.

Proposition 6.5 Liveness for AC nets. An asymmetric choice (AC) Petri net is live if and only if every siphon in the net is a controlled siphon.

Proof. See [Barkaoui and Abdallah, 1995, Barkaoui, 1995].   ∎

Remark. See section 2.4 for definitions of free choice and asymmetric choice nets.

Remark. Nonminimal siphons always contain at least one minimal siphon, so it is only necessary to examine a net's minimal siphons when applying the propositions above.

Example. The Petri net of Figure 6.2 is live. This follows from Proposition 6.5 since it is an AC net and its single minimal siphon is invariant-controlled.

There are other Petri net classes for which the condition that all siphons are controlled is sufficient for demonstrating that the Petri net is live. These include linear manufacturing lines [Minoura and Ding, 1991], and production Petri nets [Banaszak and Krogh, 1990, Xing et al., 1996]. The presence of controlled siphons is then sufficient to insure liveness for a wide variety of Petri nets, and will, at the very least, insure freedom from complete deadlock for Petri nets outside this class. Another sufficient test for liveness is presented below.

6.2.2   A Liveness Test Based on Transition Invariants

[Lautenbach and Ridder, 1994] present an algorithm for determining if a net is live, given that it contains a covering nonnegative transition invariant (T-invariant). The first two steps of the algorithm involve transformations of the of the PN graph such that the transformed net contains a minimal covering canonical T-invariant[1] while maintaining the same reachable state space as the original net. Finally, a check involving the siphons and invariants of the transformed net may prove the liveness of the original net. The steps of this algorithm are outlined below.

If a Petri net is covered by a nonnegative T-invariant, then there exists an integer vector y with strictly positive elements such that $Dy = 0$, where D is the incidence matrix of the Petri net. If the vector y contains any elements that are greater than one, then the net will be transformed so that it contains an invariant of the form $y = \begin{bmatrix} 1 & 1 & \cdots & 1 \end{bmatrix}^T$.

Transformation 1: Simplification of covering invariant.

Given a Petri net with incidence matrix D and a nonnegative covering invariant y, let y_i be the ith element of y. For all i such that $y_i = k > 1$, augment the Petri net with $k - 1$ copies of transition i (including all input and output arcs).

Remark. Transformation 1 does not alter the reachable state space of the original net, but the transformed net will be covered by a T-invariant that has a known and simple form.

Example. The Petri net of Figure 6.3a has a single minimal transition invariant

$$y = \begin{bmatrix} 1 & 1 & 2 & 1 \end{bmatrix}^T$$

Since $y_3 = 2$, the net is augmented with a single copy of t_3, as shown in Figure 6.3b. The new net contains the following minimal T-invariants

$$y_1 \;=\; \begin{bmatrix} 1 & 1 & 2 & 0 & 1 \end{bmatrix}^T$$

---

[1] An invariant is, in general, an element of the kernel of a Petri net incidence matrix. Assuming the kernel of an incidence matrix is nonempty, there will be an infinite number of invariants within the net. Canonical or minimal representations provide unique representations for the basis of an integer matrix. See [Murata, 1989, Peterson, 1981, Reisig, 1985] for details. See [Martinez and Silva, 1980] for an algorithm for computing invariants in minimal form.

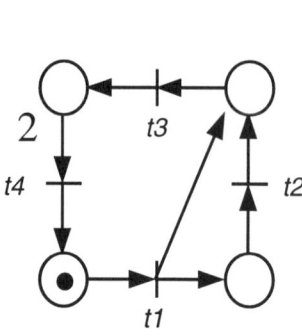

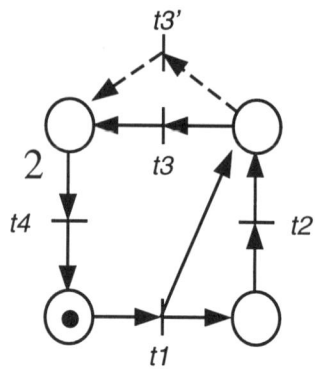

Figure 6.3a.   A PN with a covering T-invariant.

Figure 6.3b.   Transformation of the net to eliminate values greater than one from the covering invariant.

$$y_2 \; = \; \begin{bmatrix} 1 & 1 & 0 & 2 & 1 \end{bmatrix}^T$$

so the net is covered by the (nonminimal) invariant $y = \begin{bmatrix} 1 & 1 & 1 & 1 & 1 \end{bmatrix}^T$

The second transformation will add "regulation circuits" to the PN such that the net's minimal invariants are fused into a single minimal covering canonical invariant.

Transformation 2: Fusion of minimal invariants.

For all pairs of transitions $(t_i, t_j, i \neq j)$ with at least one common input place but of different minimal T-invariants, add a *regulation circuit* to the transition pair. The regulation circuit includes two new places, r_1 and r_2, such that

$$\begin{aligned}
\bullet r_1 &= t_i & \bullet r_2 &= t_j \\
r_1 \bullet &= t_j & r_2 \bullet &= t_i
\end{aligned}$$

Example. The Petri net of Figure 6.4 contains the following minimal T-invariants:

$$\begin{aligned}
y_1 &= \begin{bmatrix} 1 & 0 & 0 & 1 \end{bmatrix}^T \\
y_2 &= \begin{bmatrix} 0 & 1 & 1 & 0 \end{bmatrix}^T
\end{aligned}$$

Thus it is covered by $y = \begin{bmatrix} 1 & 1 & 1 & 1 \end{bmatrix}^T$ and it is not necessary to apply Transformation 1. The two minimal invariants will be fused into a single invariant using Transformation 2.

Transitions t_2 and t_4 are included in different invariants but are both enabled by p_2. Similarly, t_3 and t_4 are enabled by p_3. Thus regulation circuits are set up around the two pairs, $\{t_2, t_4\}$ and $\{t_3, t_4\}$, as shown in Figure 6.5. The augmented net is covered by the unique, minimal invariant $y = \begin{bmatrix} 1 & 1 & 1 & 1 \end{bmatrix}^T$.

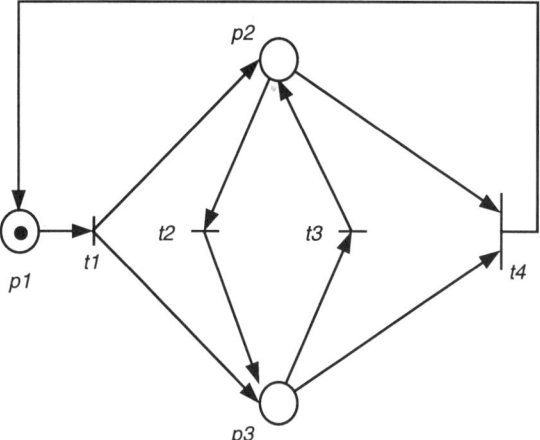

Figure 6.4.   A bounded net with a covering T-invariant.

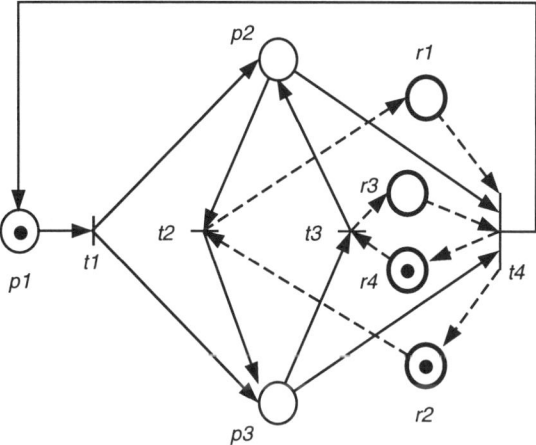

Figure 6.5.   Addition of regulation circuits to the net of Figure 6.4.

Remark. The initial conditions of the places in a regulation circuit are completely arbitrary, as will be shown below. The initial markings of r_1 through r_4 in Figure 6.5 were chosen such that t_2 and t_3 must fire before t_4 can fire, but this is not necessary.

Proposition 6.6 Sufficient condition for liveness. Given a bounded Petri net covered by a nonnegative T-invariant, apply Transformations 1 and 2 above to the net such that the transformed net is covered by a minimal canonical T-invariant. Let D' be the incidence matrix of the transformed net. The original, untransformed net is live if for all minimal siphons s of the transformed net,

$$\exists z \geq 0 \text{ s.t. } z^T D' = s^T D' \text{ and } (s - z)^T \mu_0' > 0$$

where μ_0' is the initial marking of the transformed net. The marking of the regulating places may be given any positive values in order to meet the inequality.

Proof. See [Lautenbach and Ridder, 1994]   ∎

Remark. Technically a siphon is a set of places. The variable s in Proposition 6.6 is a nonnegative integer place-vector with support equal to one of the net's minimal siphons.

Remark. A place invariant x satisfies $x^T D = 0$. Let X be a matrix whose columns form a basis of the kernel of D', then all solutions for z that satisfy $z^T D' = s^T D'$ can be written

$$z = s + Xa$$

where a is an arbitrary integer vector. In order to satisfy the conditions of Proposition 6.6, a must be chosen such that z is nonnegative. Thus the condition of the proposition can also be expressed

$$\exists a \text{ s.t. } s + Xa \geq 0 \text{ and } -a^T X^T \mu_0' > 0$$

Remark. The nonnegative invariants of a net also correspond to siphons in that net. In this case $x = s$. Let $z = 0$, so

$$z^T D' = 0 = x^T D' = s^T D'$$

In this case we need only check that $s^T \mu_0' > 0$, i.e., we need only check that the siphon is initially marked.

Example. The Petri net of Figure 6.4 is not a member of the class of free choice or asymmetric choice nets, so the propositions of 6.2.1 are not sufficient for guaranteeing that the net is live. However the net is covered by a T-invariant, so its liveness can be tested using Proposition 6.6. The transformation required by the proposition was shown in Figure 6.5. The incidence matrix of the transformed net is

$$D' = \begin{bmatrix} -1 & 0 & 0 & 1 \\ 1 & -1 & 1 & -1 \\ 1 & 1 & -1 & -1 \\ 0 & 1 & 0 & -1 \\ 0 & -1 & 0 & 1 \\ 0 & 0 & 1 & -1 \\ 0 & 0 & -1 & 1 \end{bmatrix}$$

where the last four rows of D' correspond to the regulating places r_1 through r_4. Four vectors form the basis for the place invariants of this net:

$$x_1 = \begin{bmatrix} 2 & 1 & 1 & 0 & 0 & 0 & 0 \end{bmatrix}^T$$

$$x_2 = \begin{bmatrix} 0 & 0 & 0 & 1 & 1 & 0 & 0 \end{bmatrix}^T$$

$$x_3 = \begin{bmatrix} 0 & 0 & 0 & 0 & 0 & 1 & 1 \end{bmatrix}^T$$

$$x_4 = \begin{bmatrix} 1 & 1 & 0 & 1 & 0 & 0 & 1 \end{bmatrix}^T$$

Each of these vectors also corresponds to a siphon, and all of these siphons are marked. Using the algorithm from section 2.3 we find the net contains two more minimal siphons that are not equal to place invariants:

$$s_1 = \begin{bmatrix} 1 & 1 & 0 & 0 & 0 & 0 & 1 \end{bmatrix}^T$$

$$s_2 = \begin{bmatrix} 1 & 0 & 1 & 0 & 1 & 0 & 0 \end{bmatrix}^T$$

For s_1,

$$z = s_1 - (x_4 - x_2) = \begin{bmatrix} 0 & 0 & 0 & 0 & 1 & 0 & 0 \end{bmatrix}^T \geq 0$$

Because z is equal to s_1 plus elements of the kernel of D', we have $z^T D' = s_1^T D'$. The inequality from the proposition is now verified:

$$(s_1 - z)^T \mu_0' = (x_4 - x_2)^T \mu_0' =$$

$$\begin{bmatrix} 1 & 1 & 0 & 0 & -1 & 0 & 1 \end{bmatrix} \begin{bmatrix} 1 \\ 0 \\ 0 \\ 0 \\ 1 \\ 0 \\ 1 \end{bmatrix} = 1 > 0$$

Note that it is possible to choose any positive values for the last four elements of μ_0' in order to meet the inequality.

Siphon s_2 is handled analogously:

$$z = s_2 - (x_1 + x_2 - x_4) = \begin{bmatrix} 0 & 0 & 0 & 0 & 0 & 0 & 1 \end{bmatrix}^T \geq 0$$

$$(s_2 - z)^T \mu_0' = (x_1 + x_2 - x_4)^T \mu_0' =$$

$$\begin{bmatrix} 1 & 0 & 1 & 0 & 1 & 0 & -1 \end{bmatrix} \begin{bmatrix} 1 \\ 0 \\ 0 \\ 0 \\ 1 \\ 0 \\ 1 \end{bmatrix} = 1 > 0$$

All the net's siphons meet the conditions of Proposition 6.6, so the net is live.

6.3   Deadlock Avoidance

6.3.1   Deadlock Avoidance in Manufacturing Systems

The specific structures of particular applications can lend themselves to the development of specialized methods for achieving desired plant performance, including deadlock avoidance. The work of [Bogdan and Lewis, 1997, Lewis et al., 1995, Huang et al., 1995, Tacconi et al., 1996, Huang et al., 1996] is driven by the function and needs of manufacturing systems. Deadlock avoidance involves a search for the critical siphons of the net and the enforcement of linear inequalities based, in part, on this search. The work of Lewis et al. integrates techniques from Petri net theory, industrial engineering and the factory automation literature, as well as some novel techniques, to develop an approach to the control of factory/assembly related DES's.

The design method takes assembly operation specifications, e.g., which resources are required to complete which jobs, and which jobs rely on the completion of other jobs, etc., and produces a sequencer/controller with the firing dynamics of a Petri net. Valid firings are obtained based on system structure, deadlock avoidance, and optimization criteria, e.g., maximum throughput. These three criteria are implemented through boolean logic in a rule based production engine. The resulting controller is a hierarchical intelligent controller that uses a maximally permissive supervisor to avoid deadlock and a higher level controller to direct the firings based on the desired performance goal.

The work in [Huang et al., 1995, Huang et al., 1996, Lewis et al., 1995] on deadlock avoidance only applies to a specific class of nets that are relevant to the manufacturing concerns the authors are studying. The assumptions made on the plant are true for most manufacturing systems, these include the properties of "nonpreemtion, mutual exclusion, and hold-while-waiting" (see references for details). The structures and concepts in the following definitions are used in the deadlock avoidance procedure presented below.

Definition 6.7 The system's operation involves the use of a number of finite **resources**. Let r_i represent the ith resource.   ∎

Definition 6.8 If the release of resource r_i is conditional on the availability of r_j, then r_i **waits** for r_j, or $r_i \hookrightarrow r_j$. If r_i can be released immediately upon the availability of r_j, then there is a **first order wait** relation between r_i and r_j: $r_i \to r_j$.   ∎

Remark. The wait relation is transitive, e.g., if $r_1 \to r_2 \to r_3 \to r_4$, then $r_1 \hookrightarrow r_4$.

Definition 6.9 A **circular wait** (CW) is a set of resources that all wait on each other. A **simple circular wait** is described as $r_1 \to r_2 \to \ldots \to r_q \to r_1$. A set of resources C is a **general circular wait** if $\forall r_i, r_j \in C, r_i \hookrightarrow r_j$.   ∎

Remark. All circular waits involve at least one shared resource. The assumptions made on the plant indicate that deadlock can only occur if the plant contains a circular wait.

Each CW is associated with a "critical siphon" that includes the places that make up the CW. An empty CW may not be blocked, but if its associated siphon is empty, then it will be. A critical siphon indicates all the places that must become empty in order for a CW to be blocked.

Definition 6.10 A **critical subsystem** is a minimal covering of the places of a CW by a place invariant.   ∎

Remark. In this work, every resource loop is a place invariant, i.e., a resource together with its job set forms a place invariant. Note that this is also true when finite resources are modeled using invariant based controllers as described in section 6.1.

Deadlock is avoided through the following design procedure.

1. Find the place invariants in the net.

2. find all the circular waits (loops in the net that use a finite resource).

3. Find all of the critical siphons (one is associated with each circular wait).

4. Find all of the critical subsystems (a minimal covering over a critical siphon by a place invariant).

5. For each critical subsystem S associated with CW c, enforce

$$\sum_{p_i \in S} \mu_i < c^T \mu_{p0} \tag{6.2}$$

In words, inequality (6.2) indicates that the total number of jobs-in-progress in a given subsystem must be less than the total number of resources in the associated CW. If this inequality can be enforced on all subsystems, then deadlock will be avoided.

Procedures are provided by the authors for finding all of the structures in steps 1) through 4). The constraints that are to be enforced in 5) are in the spirit of supervisory control, i.e., they are maximally permissive constraints and may be enforced using the methods of chapter 3. Another level of the controller is used to make decisions as to which transitions are to be fired (when there is a choice) based on its own optimization criteria and assuming that the supervisor is allowing the desired event to occur.

The work of [Xing et al., 1996] deals with another PN model motivated by manufacturing concerns called the *production Petri net* (PPN). A PPN includes strings of places, representing steps in one or more processes, with resource usage constraints imposed on them. PPN's can be created using the technique of section 6.1.

A deadlock state is reachable in a PPN when it contains a *deadlock structure*, which is similar to a siphon but is defined specifically for PPN's on sets of transitions rather than places; see [Xing et al., 1996] for details. Deadlock avoidance policies must insure that

$$\sum_{p_i \in S_j} \mu_i < r_j \tag{6.3}$$

where S_j is the the set of plant places associated with deadlock structure j, and r_j is the number of resources associated with deadlock structure j. Note the similarity of (6.2) and (6.3) in both form and meaning.

Methods are provided in [Xing et al., 1996] for creating deadlock-avoiding Petri net supervisors. The supervisors are shown to be maximally permissive when there is more than one of each available resource. These supervisors are equivalent to the invariant based controllers that would result from the enforcement of the indicated linear inequalities.

The deadlock avoidance technique of the following section presents a less application specific approach to the problem.

6.3.2   Deadlock Avoidance through Supervisory Control

Siphon Controlling Supervisors

A supervisory control technique is introduced in [Barkaoui and Abdallah, 1995] for handling the problem when not all of the siphons in a given Petri net are controlled, i.e., when certain siphons within the net may lose all of their tokens and thus produce local deadlocks. The method involves adding a place for each uncontrolled siphon in the net such that they become controlled. These controller places act to restrict behaviors in the original plant that would lead to deadlock, thus they play the part of a supervisory controller, allowing the plant's state to evolve unrestricted except to prevent transition firings that lead to "forbidden states." An outline of this technique is described below.

Definition 6.11 A **conservative** Petri net is covered by a nonnegative place invariant.
∎

Remark. Clearly a conservative PN is bounded, though a bounded PN is not necessarily conservative.

Definition 6.12 All of the siphons of a **well-marked** Petri net are initially marked, i.e., each siphon starts with at least one token.
∎

Remark. A net that is not well-marked can not be live. The places of an unmarked siphon will never become marked, thus all of the transitions to which the siphon's places direct their arcs are dead.

The deadlock-avoidance procedure in [Barkaoui and Abdallah, 1995] follows. Given a conservative, well-marked Petri net with uncontrolled siphons, for each uncontrolled siphon S, create a control place c such that

$$
\begin{aligned}
c\bullet &= \{t \in S\bullet : |\bullet t \cap S| > |t \bullet \cap S|\} \\
\bullet c &= \{t \in \bullet S : |t \bullet \cap S| > |\bullet t \cap S|\}
\end{aligned}
\tag{6.4}
$$

where the notation $|x|$ refers to the number of elements in the set x, and the weights of the arc transitions are given by the differences $|\bullet t \cap S| - |t \bullet \cap S|$ and $|t \bullet \cap S| - |\bullet t \cap S|$ for the controller place's output and input arcs respectively. The initial marking of the control place, μ_{c_0}, is given by

$$
\mu_{c_0} = \sum_{p_i \in S} \mu_{i_0} - 1
\tag{6.5}
$$

where μ_{i_0} is the initial marking of place p_i in the plant.

Each control place insures that its siphon will never be emptied of all of its tokens. An analysis of the synthesis technique, (6.4) and (6.5), shows that this is done by creating place invariants in the controlled Petri net. For each control place c, associated with siphon S, the following place invariant is established in the controlled Petri net.

$$
\sum_{p_i \in S} \mu_i - \mu_c = 1
\tag{6.6}
$$

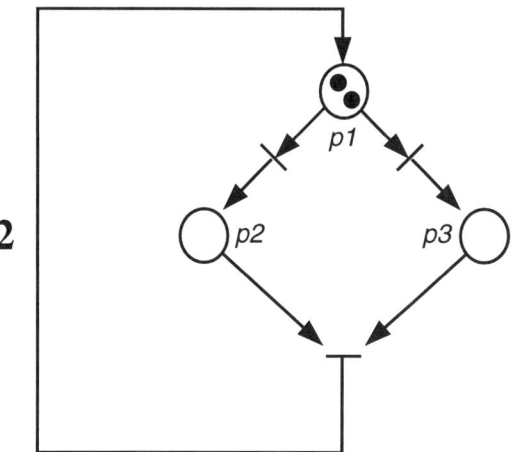

Figure 6.6.   A free choice Petri net with uncontrolled siphons.

Thus the synthesis technique causes formerly uncontrolled siphons in the plant to become invariant-controlled siphons in the closed loop system.

Example.   The free choice Petri net of Figure 6.6 is conservative: it contains a covering place invariant $x = [1\ 1\ 1]^T$. The three places form a trap, and this trap is minimal. The net contains three siphons:

$$
\begin{aligned}
S_1 &= \{p_1, p_2\} \\
S_2 &= \{p_1, p_3\} \\
S_3 &= \{p_1, p_2, p_3\}
\end{aligned}
$$

Place p_1 is marked with two tokens, and it is involved in all three of the net's siphons, thus the net is well-marked.

Siphon S_3 is both trap and invariant-controlled, however it is not minimal. Neither minimal siphon, S_1 nor S_2, is controlled. It is easy to see that deadlock would result if the two tokens in p_1 were to both transfer to p_2 or p_3.

A deadlock-avoiding controller is constructed according to (6.4) and (6.5). Control places c_1 and c_2 are associated with siphons S_1 and S_2 respectively. All of the siphons in the closed-loop system, shown in Figure 6.7, are controlled, and the net is live. Note that the resulting net is no longer a free choice net, though it is asymmetric choice.

Controlling all of the formerly uncontrolled siphons in a net is sufficient for insuring liveness for a class of Petri nets. However, as pointed out by [Ezpeleta et al., 1995], the act of adding supervisors to control siphons may actually introduce new uncontrolled siphons into the net, allowing the closed loop system to become deadlocked. The designer should recheck the siphons of the closed loop system as a whole after using this procedure to verify that the system is indeed deadlock-free. If the system is not, it may be possible iterate and create supervisors for controlling the new uncontrolled siphons, or the designer may wish to try a different technique, such as that in [Ezpeleta et al., 1995].

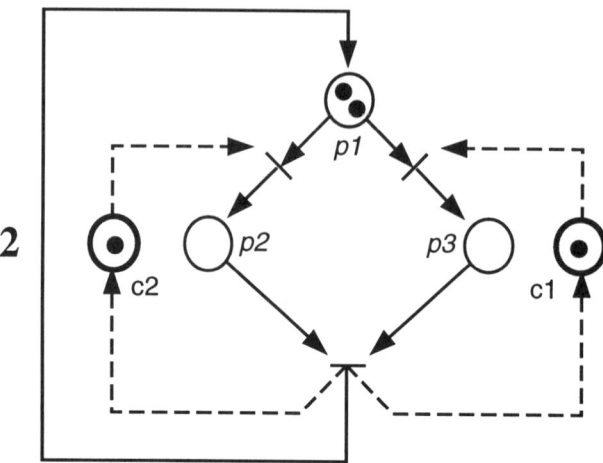

Figure 6.7. The controller, shown with bold places and dashed arcs, insures the liveness of the FC net of Figure 6.6.

For nets that can not be made live using siphon controlling supervisors, the following proposition is presented in [Barkaoui and Abdallah, 1995].

Proposition 6.13 Loss of liveness in controlled nets. If the siphons in a Petri net are controlled using (6.4) and (6.5), but the net is not live (at least one of the transitions has become "dead"), then the marking of at least one of the control places has become zero.

Proof. See [Barkaoui and Abdallah, 1995]. ∎

Based on this proposition, an algorithm appears in [Barkaoui and Abdallah, 1995] that determines which transitions should fire in order to cause the filling of control places in the fewest number of steps. Because this algorithm actively seeks transitions that should fire, rather than the simple enabling and disabling of transitions of a supervisory controller, it is not discussed here.

Relation to Invariant Based Control

The controllers of Barkaoui and Abdallah enforce the invariant equation (6.6). This is equivalent to enforcing the following inequality

$$\sum_{p_i \in S} \mu_i \geq 1 \qquad (6.7)$$

where μ_c plays the part of a (nonnegative) excess variable. The inequality is intuitively appealing, simply stating that the siphon should never be emptied of tokens. This is the primary consideration of the deadlock and liveness conditions of Propositions 6.2 through 6.5.

Inequality (6.7) can be enforced using an invariant based supervisor. Framing the problem this way extends the deadlock avoidance method of [Barkaoui and Abdallah, 1995] to the control procedures of this book. Equivalently, it adds the ability to handle a variety of other supervisory constraints as well as uncontrollable and unobservable transitions to the method of [Barkaoui and Abdallah, 1995].

As usual, the form of an individual constraint imposed on the plant state is

$$l^T \mu_p \leq b \qquad (6.8)$$

For inequality (6.7),

$$l_i = \begin{cases} -1 & \text{if } p_i \in S \\ 0 & \text{else} \end{cases} \qquad (6.9)$$
$$b = -1$$

for $i = 1 \ldots n$, where l_i is the ith element of l. The definition of (6.9) shows that both sides of inequality (6.7) were multiplied by -1 to achieve the "less-than-or-equal-to" form of inequality (6.8).

The use of linear inequality constraints to prevent deadlock is illustrated in section 8.1.2 for the "cat and mouse" problem. The "unreliable machine" problem of section 8.3 shows how safety and resource constraints, uncontrollable transitions, and deadlock avoidance can all be handled in a unified manner with the invariant based control method.

Existence of Liveness-Enforcing Supervisors

Some nets can not be made live by any supervisory controller. For example, there is clearly no transition enabling/disabling supervisory control law that will make the net of Figure 6.8a live.

Figure 6.8a. A net for which no liveness-enforcing controller exists.

Figure 6.8b. A siphon controlling supervisor is added to the net.

[Sreenivas, 1997b, Sreenivas, 1997a] has studied the existence of liveness-enforcing supervisors. The existence of such a supervisor depends on the structure of the net, the initial conditions, and the uncontrollable transitions. A necessary and sufficient condition for the existence of a static state feedback liveness-enforcing supervisor appears in [Sreenivas, 1997b]. The problem of testing this condition is decidable when the net is bounded, or all of its transitions are controllable, otherwise it is undecidable.

A theorem in [Sreenivas, 1997a] states that a static state feedback supervisory policy will enforce liveness in a free choice net if and only if all of the plant's siphons remain marked for all reachable markings, *and the supervisor always permits some transition to fire.* The invariant based controller shown in Figure 6.8b insures that the plant's single siphon will never be emptied, however it does not permit any transition to fire, so it is not a liveness-enforcing supervisor.

The controllers considered in [Sreenivas, 1997b, Sreenivas, 1997a] are static maps, indicating whether every controllable transition is enabled or disabled based on the current state. Invariant based controllers are dynamic but also enforce static state feedback laws (see Proposition 7.6 of section 7.5). The PN representation of the supervisor is advantageous when it comes time to analyze the closed loop system. For example, analysis of the controlled system in Figure 6.8b shows that the addition of the supervisor also introduced a new uncontrolled siphon. This siphon is not only uncontrolled, but it is also unmarked. The closed loop system is not well marked, thus the supervisor will not enforce liveness. Siphon-based analysis of the closed loop system can not be performed when the controller is modeled by some external function rather than a structurally integrated and connected Petri net.

The analysis above is consistent with the result in [Sreenivas, 1997b] that states that the existence of a liveness-enforcing supervisor is a decidable problem for a bounded Petri net. Suppose, given the net in Figure 6.8b, we decide to add another control place to regulate the newly created uncontrolled siphon. This new constraint will state that the number of tokens in the first control place should be greater than or equal to one. However, the initial number of tokens in this place was already determined to be zero. This means that the new constraint will violate condition (3.9) of Theorem 3.2, and there is no supervisor (invariant based or otherwise) that will solve the problem. As long as the number of tokens in the plant of Figure 6.8a is finite, the iterative addition of siphon controlling supervisors will finally end in a situation that violates condition (3.9).

7 OTHER CONTROL SPECIFICATIONS

Constraints of the form $L\mu_p \leq b$ (inequality (3.4)) are useful for representing a large variety of forbidden state problems, even when the initial specification does not appear to directly involve the states of system. This chapter will show how several common varieties of system constraints can be written in the form of (3.4), enabling the use of invariant based control.

Constraints need not have an upper bound on the range of feasible μ_p solutions. For example, some systems may require resource reserve constraints. Consider a multiprocessor computer with processor allocation modeled by a Petri net. One constraint on the system might be that two processors must always be available to handle user I/O. The constraint could be written $\mu_i \geq 2$, which is equivalent to $-\mu_i \leq -2$ and is in the form of inequality (3.4).

Section 7.1 presents a brief discussion of equality constraints, $L\mu_p = b$. It is shown that attempts to enforce these with invariant based controllers can lead to deadlock.

Constraints may be written in terms of events rather than plant states. For example, one might wish to prevent the simultaneous firing of two transitions. This constraint would be written $q_i + q_j \leq 1$, i.e., in terms of the firing vector, rather than the plant marking. A direct enforcement of this constraint would involve using a supervisor to disable one of these transitions whenever the other transition fires. The constraint could also be enforced indirectly by using a supervisor to eliminate states that would leave both of the transitions simultaneously enabled. Section 7.2 explains how both direct and indirect enforcement of these constraints can be handled using the invariant based control method.

In section 7.3, a class of logical predicates on plant behavior are transformed into systems of linear inequalities to be enforced by a supervisor. Section 7.4 explains how the techniques for supervision of ordinary Petri nets can be expanded to timed Petri nets. This section includes techniques for enforcing constraints that make reference to time. The chapter concludes with a look at the limits of the constraint inequality in section 7.5.

7.1   Equality Constraints

Equality constraints have the form

$$L\mu_p = b \tag{7.1}$$

Equation (7.1) defines place invariants on the original process net. This is really a specification for the system and should have been incorporated into the Petri net model before attempting to use supervisory control. If this invariant is not already part of the Petri net model, it should become one by modifying the incidence matrix D_p of the plant so that equation (2.8) holds, where $x^T = L$ in equation (7.1). The new elements of D_p represent the arcs that should be added to the Petri net so that the place invariants are enforced.

It may seem feasible to use the place invariant control method to force $L\mu_p \leq b$ and $L\mu_p \geq b$ in order to achieve the constraint of equation (7.1). Unfortunately this approach will produce undesirable results as described by the following proposition.

Proposition 7.1   Enforcement of equality constraints leads to deadlock. Enforcing constraint (7.1) by creating invariant based controllers for the constraints

$$L\mu_p \leq b, \text{ and} \tag{7.2}$$
$$-L\mu_p \leq -b \tag{7.3}$$

will

1. have no effect on the plant's behavior, or

2. create a local deadlock in the plant (the system will not be live).

Proof. Suppose that the natural behavior of the plant already meets the desired constraint. In this case, L describes a set of place invariants in the plant and $LD_p = 0$. Equation (3.10) shows that the controller for constraint (7.2) or (7.3) is given by $D_c = \pm LD_p = 0$. Thus the controller will have no arcs to the plant transitions, and it will have no effect on the plant's behavior.

Now suppose that L does not include natural invariants of the plant. In this case, the controller incidence matrices for (7.1) and (7.2) are given by

$$D_{c1} = -LD_p \neq 0$$
$$D_{c2} = LD_p = -D_{c1}$$

Since $D_{c1} = -D_{c2}$, all output arcs of the places in D_{c1} are input arcs of the places in D_{c2} and vice versa. Thus the set of control places forms a siphon.

The initial marking, μ_{p_0}, of the plant must satisfy $L\mu_{p_0} = b$ or it would not have been feasible to attempt to enforce (7.1). Equation (3.11) gives the initial markings of the control places:

$$
\begin{aligned}
\mu_{c1_0} &= b - L\mu_{p_0} = 0 \\
\mu_{c2_0} &= -b + L\mu_{p_0} = 0
\end{aligned}
$$

Thus the set of control places forms an unmarked siphon and all of the transitions to which these places are connected will be dead according to Proposition 6.2.   ■

7.2   Constraints involving the Firing Vector

Certain control goals may involve the firing vector of the Petri net as well as or opposed to the places. For example one might need to insure that two transitions do not fire simultaneously or that a certain transition is never allowed to fire when a certain place holds a token. As mentioned in the introduction to this chapter, there are two ways that constraints like these may be viewed. For the constraint

$$\mu_i + q_j \leq 1 \tag{7.4}$$

do we mean that transition j should be disabled whenever place i contains a token, or do we mean that all plant states that would allow transition j to be enabled are forbidden whenever place i contains a token? The answer to this question lies in the particulars of a given plant and its operation. Both means of enforcing the constraint can be useful for different problems.

Section 7.2.1 describes rules for enforcing firing vector constraints using the "direct" interpretation, i.e., transitions are explicitly disabled in order to satisfy the inequality. Algebraic schemes for handling the "indirect" interpretation of firing vector constraints were proposed in [Yamalidou et al., 1996]. A new approach is presented in section 7.2.2 that uses the concept of uncontrollable transitions to force a correct interpretation of each constraint, thus avoiding the enumeration of separate cases that appeared in [Yamalidou et al., 1996].

7.2.1   Direct Realization

Assume that the plant must satisfy constraint (7.4). The direct interpretation of this constraint implies that transition t_j cannot fire if place p_i is marked, and, of course, place i can never contain more than one token. The invariant based control technique requires that all variables in a constraint inequality be members of a Petri net marking vector. To bring this constraint to a form that contains elements of the marking vector only, the plant structure is temporarily transformed. Transition j is replaced by two transitions and a place between them, as shown in Figure 7.1. This transformation is artificial and will not effect the Petri net model of the plant. *Its sole purpose is to introduce the place p_j', which records the firing of the transition t_j.* After the controller has been computed the plant will be transformed back to its original form.

The marking μ_j' of p_j' replaces q_j in constraint (7.4), which becomes

$$\mu_i + \mu_j' \leq 1 \tag{7.5}$$

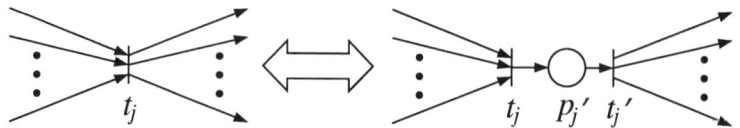

Figure 7.1.   Transformation of a transition.

The constraint now contains only μ's and the controller can be computed using the place invariant control method. Since the method produces a controller consisting of places and arcs only, no part of the controller is connected directly to the place $p_j{}'$ of the transformation. After the controller structure is computed, the two transitions and the place of the transformation collapse to the original transition thus restoring the original form of the plant while maintaining the enforcement of the new constraint. The same transformation is done to all the transitions that appear in the constraints. Those constraints that contain only q's are treated in the same way.

In terms of Figure 7.1, output arcs from the controller would normally be connected to transition t_j, and input arcs to the controller would be connected to t'_j. The act of collapsing the transformed structure back to its original form will cause both the input and output arcs to be connected to the original transition t_j. *This means that the controller will contain self loops to the transitions indicated in the constraints.* Separate D_c^+ and D_c^- matrices must be maintained for the controller, and enabling rule (2.1) must be used to check the feasibility of firing vectors.

In summary, given a plant (D_p, μ_{p_0}) and constraint

$$l^T \mu_p + f^T q \le b, f \ge 0 \tag{7.6}$$

The invariant based controller $(D_c = D_c^+ - D_c^-, \mu_{c0})$ is given by

$$
\begin{aligned}
D_c^+ &= D_{lc}^+ + D_{fc}^+ \\
D_c^- &= D_{lc}^- + D_{fc}^- \\
\mu_{c0} &= b - l^T \mu_{p0}
\end{aligned}
$$

where

$$D_{fc}^+ = D_{fc}^- = f \tag{7.7}$$

and

$$D_{lc}^+(j) = \begin{cases} D_{lc}(j) & \text{if } D_{lc}(j) > 0 \\ 0 & \text{else} \end{cases} \tag{7.8}$$

$$D_{lc}^-(j) = \begin{cases} -D_{lc}(j) & \text{if } D_{lc} < 0 \\ 0 & \text{else} \end{cases} \tag{7.9}$$

where

$$D_{lc} = -l^T D_p \tag{7.10}$$

and $D_{lc}(j)$ is the jth element of the vector D_{lc}. Examples involving an asynchronous transfer mode switch and a continuous/discrete hybrid system in sections 8.5 and 8.7 illustrate the enforcement of direct firing vector constraints.

The remainder of this section provides an analysis of the admissibility of firing vector constraints using the direct interpretation. Similar to Corollary 4.6, the following corollary defines when a constraint on the firing vector of a Petri net is admissible.

Corollary 7.2   Transition-constraint admissibility.   A single vector constraint $f^T q \leq b$, where $f, b \geq 0$, is *admissible* under direct transition-constraint implementation on a plant with controllable transitions T_c, if $\forall j$ s.t. $f_j \neq 0, t_j \in T_c$.

Proof. The proof is by Proposition 4.5 on general constraint controllability. The direct transition-constraint enforcement method for the constraint $f^T q \leq b$ is maximally permissive since it is constructed as an invariant based controller. The initial marking of the controller $\mu_{c_0} = b$ is valid if $b \geq 0$. The incidence matrix of the controller $D_c^+ = D_c^- = f^T$ contains input arcs to all transitions j such that $f_j \neq 0$. If all of these transitions are controllable, then the controller draws no arcs to uncontrollable transitions and the constraint is admissible.   ∎

The admissibility of combined marking/firing constraints, $l^T \mu_p + f^T q \leq b$, requires the following definition.

Definition 7.3   A constraint of the form (7.6) is called **uncoupled** if

$$T_l \cap T_f = \emptyset$$

where T_l is the set of transitions that are connected to the controller induced by the $l^T \mu_p$ portion of the constraint (transitions t_j such that $D_{lc}(j) \neq 0$ in equation (7.10)), and T_f is the set of transitions connected to the controller induced by the $f^T q$ portion of the constraint (transitions t_j such that $f_j \neq 0$).   ∎

Constraint (7.6) is uncoupled if the transitions involved in the $l^T \mu_p$ and $f^T q$ portions of the constraint are mutually exclusive.

Proposition 7.4   Uncoupled place/transition constraints.   A vector constraint of form (7.6) is uncoupled iff

$$\forall i \text{ s.t. } f_i \neq 0, l^T D_p e_i = 0 \tag{7.11}$$

where e_i is a zero-vector with a 1 in the ith place.

Proof. The set of plant transitions that will contain arcs to or from the controller is determined from the controller synthesis equations. This set is the union of the transitions connected by arcs induced by the $l^T \mu_p$ and $f^T q$ portions of the constraint, i.e., $T_f \cup T_l$. Equation (7.7) indicates that

$$T_f = \{t_j | f_j \neq 0\} \tag{7.12}$$

and equations (7.8) and (7.9) show

$$T_l = \{t_j | l^T D_p e_j \neq 0\} \tag{7.13}$$

Combining these with condition (7.11) implies

$$T_l \cap T_f = \emptyset \tag{7.14}$$

The sets of transitions used by the two portions of the controller are mutually exclusive and the constraint is uncoupled according to definition 7.3.

It is easy to see that if the constraints are uncoupled, i.e. $T_l \cap T_f = \emptyset$, then (7.11) must be true by working backward through the development above. If (7.11) were not true, then there would exist some $t_i \in T_f$ and $t_i \in T_l$, which would imply through equations (7.12) and (7.13) that $T_l \cap T_f \neq \emptyset$ and the constraints were coupled.   ∎

Proposition 7.5 Place/transition constraint admissibility. An uncoupled vector constraint of form (7.6) is to be imposed on a plant (D_p, μ_{p_0}) with uncontrollable transitions T_{uc} and controllable transitions $T_c, T_{uc} \cap T_c = \emptyset$.
    if the constraints

$$l^T \mu_p \leq b \tag{7.15}$$
$$f^T q \leq |b| \tag{7.16}$$

are both admissible then $l^T \mu_p + f^T q \leq b$ is admissible.

Proof. If the admissibility of constraints (7.15) and (7.16) imply that (7.6) is admissible, then the inadmissibility of (7.6) will imply that either (7.15) or (7.16) is inadmissible or both. For $l^T \mu_p + f^T q \leq b$ to be inadmissible, it must lie outside the range of the plant's initial conditions, or a maximally permissive controller that enforces the constraint would attempt to inhibit an otherwise enabled transition in the set T_{uc}. Because (7.6) is uncoupled, the transitions that are connected to the controller places, T_f and T_l, are mutually exclusive. This means that at least one of the following three cases must be true for $l^T \mu_p + f^T q \leq b$ to be inadmissible.

1. The initial conditions of the plant violate the constraint.

2. The controller would attempt to inhibit a transition $t_j \in T_{uc}$, where $t_j \in T_f$.

3. Or the controller would attempt to inhibit an otherwise enabled transition $t_j \in T_{uc}$, where $t_j \in T_l$.

Case 1: The initial state of the plant is μ_{p_0}. The firings indicated by the vector q are determined after the system commences its run, thus if the initial conditions of the plant violate constraint (7.6), then

$$l^T \mu_{p_0} > b$$

This condition would also indicate that the constraint $l^T \mu_p \leq b$ is inadmissible according to Corollary 4.6.

Case 2: According to the construction of the maximally permissive controller for direct transition constraints, the transitions in the set T_f are identical to the transitions that receive controller arcs in the constraint $f^T q \leq |b|$. If the controller attempts to inhibit an uncontrollable transition in this set, then the constraint $f^T q \leq |b|$ is inadmissible according to Corollary 7.2.

Case 3: The construction of the maximally permissive controller for the constraint $l^T \mu_p \leq b$ shows that the transitions that receive controller arcs for this constraint are identical to the set T_l. If the controller for constraint (7.6) attempts to disable an otherwise enabled transition in the set T_l, then the constraint $l^T \mu_p \leq b$ will be inadmissible according to Corollary 4.6.

Thus if both $l^T \mu_p \leq b$ and $f^T q \leq |b|$ are admissible, then $l^T \mu_p + f^T q \leq b$ is also admissible.   ■

The use of the propositions and definitions above are illustrated in the "three tanks" example of section 8.6.2.

7.2.2   Indirect Realization

Firing vector constraints can be realized by preventing the states that would allow the undesirable transition firing; this situation is analogous to the case when a transition is uncontrollable but is involved with regular marking constraints. Illegal states are prevented in the presence of uncontrollable transitions by preventing those states that could lead, through uncontrollable transitions, to the explicitly forbidden states. The results for uncontrollable transitions can be applied to constraints involving the firing vector through utilization of the graph transformations discussed in the previous section.

The procedure is illustrated in the example below.

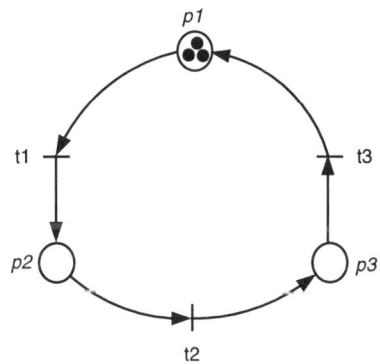
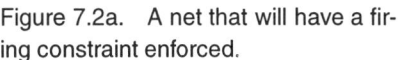
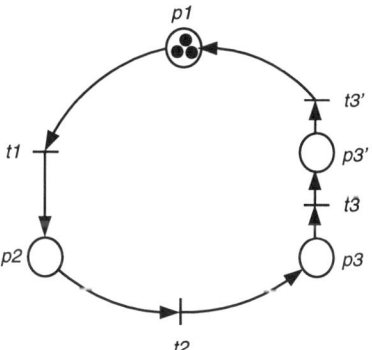

Figure 7.2a.   A net that will have a firing constraint enforced.

Figure   7.2b. The graph-transformation introduces a place to record the firing of t_3.

Example. For the plant of Figure 7.2a, we wish to enforce the constraint

$$\mu_2 + q_3 \leq 1 \qquad (7.17)$$

Place p_2 is never to have more than one token and transition t_3 should never fire when this place is occupied. This problem could be solved simply by applying the technique of the previous section, but suppose instead of directly controlling the transition we want to prevent the states that could lead to the constraint being violated. Because the Petri net is so simple, we can see by inspection that the job can be done by enforcing the constraint $\mu_2 + \mu_3 \leq 1$. But how can this new constraint be generated automatically based on (7.17)?

Suppose we perform the graph transformation on this net as shown in Figure 7.2b. The transformation changes (7.17) to

$$\mu_2 + \mu_3' \leq 1 \tag{7.18}$$

If we continue to follow the procedure described in section 7.2.1, we would end up with a controller that directly enables and disables transition t_3. In order to prevent this from occurring, we will label transition t_3 as uncontrollable and then continue with the procedure.

Applying the method for constraint transformations in the presence of uncontrollable transitions from chapter 5 to (7.18), we obtain the following transformed constraint:

$$\mu_2 + \mu_3 + \mu_3' \leq 1 \tag{7.19}$$

The controller that enforces this constraint can be automatically generated using the place invariant method and is shown in Figure 7.3a.

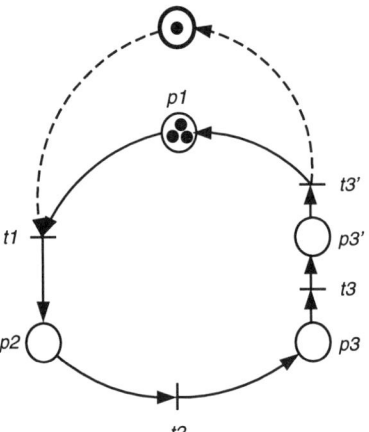
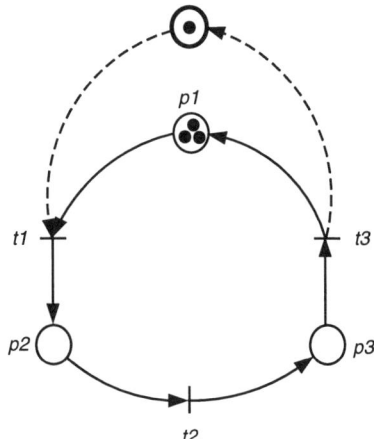

Figure 7.3a. The net of Figure 7.2b with its Petri net controller.

Figure 7.3b. The dummy place is re-moved in the final step.

The final stage is then to collapse the controlled net back to the form it had before the graph transformation was performed. The final controlled version of the net is shown in Figure 7.3b. Transition t_3 will not fire when place p_2 contains a token because the controller only allows one token at a time in places p_2 and p_3, which is the desired result.

The procedure used in the example is summarized below. Given a constraint

$$l^T \mu_p + f^T q \leq b \tag{7.20}$$

where l may be zero, indicating a constraint on the firing vector alone, first perform a transformation of the plant such that each transition specified by a nonzero entry in f includes a dummy place to mark its firing as described in section 7.2.1. The marking vector μ' is associated with the dummy places and the constraint becomes

$$\begin{bmatrix} l^T & f^T \end{bmatrix} \begin{bmatrix} \mu_p \\ \mu' \end{bmatrix} \leq b \qquad (7.21)$$

Next mark all transitions specified by nonzero entries in f as uncontrollable. Use the techniques of chapter 5 to find an admissible constraint that enforces the inadmissible constraint (7.21) and construct a supervising controller for this constraint. This will have the effect of preventing the states that could lead to (7.21) being violated. It will prevent the transitions specified by f from being enabled such that constraint (7.20) could be violated.

Finally, collapse the net back to its original form by removing the dummy places and extra transitions as described in section 7.2.1. A more complex example of this procedure is given in section 8.2 for the "automated guided vehicle."

7.3   Logical Constraints on System Behavior

The transformation of logic-based constraints on system behavior into systems of linear inequalities has been studied by [Yamalidou, 1991, Yamalidou and Kantor, 1991]. These transformations apply to safe nets, meaning that no place in the network can have more than one token at any time. In this case, all places have two states: either they contain a token or they do not. Similarly all transitions can be viewed as having two states: either they will fire in the current iteration of the system's evolution or they will not. This means that both places and transitions have binary valued states in a safe net and they can be viewed as boolean variables.

Consider the network in Figure 7.4a. We wish to enforce the constraint

$$\text{if } \mu_1 \neq 0, \text{ then } q_3 = 0 \qquad (7.22)$$

One method of doing this would be to introduce an inhibitor arc into the Petri net model as shown in Figure 7.4b. The arc between p_1 and t_3 is terminated with a circle indicating that the arc will inhibit the firing of transition 3 whenever place 1 contains any tokens. Unfortunately, with the addition of an inhibitor arc, we are no longer dealing with an ordinary Petri net and have lost some of the ability to analyze the net in a convenient manner due to the increased complexity of the transition enabling rule.

If we assume that the net is safe, constraint (7.22) can be implemented by transforming it into a linear inequality:

$$\mu_1 + q_3 \leq 1 \qquad (7.23)$$

The controller for enforcing this linear inequality is shown in Figure 7.4c. It was constructed using the direct method for implementation of firing vector constraints (section 7.2.1).

If the network in Figure 7.4a were not safe, i.e., transition 1 could fire multiple times allowing tokens to pile up in p_1, then it is still true that the controller in Figure

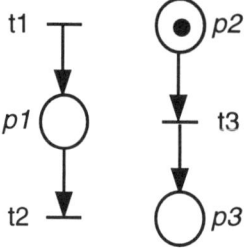

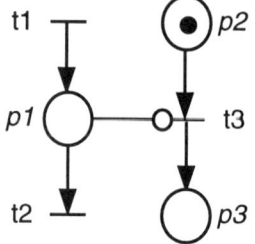

Figure 7.4a.   Constraint (7.22) is to be enforced on this network.

Figure 7.4b.   Method 1 uses an inhibitor arc.

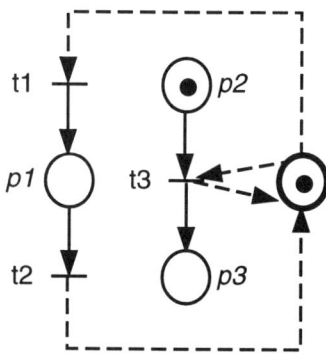

Figure 7.4c.   Method 2 with an invariant based controller.

7.4c would enforce constraint (7.22). Unfortunately it also has the secondary effect of preventing p_1 from ever containing more than one token. This is due to the literal interpretation of inequality (7.23) and thus, if the net were not safe, constraint (7.23) would be more restrictive than constraint (7.22) and the supervisor could not be considered maximally permissive.

A formal procedure for translating logical implications into linear inequalities appropriate for use as constraints on Petri net behavior appears in [Yamalidou, 1991] (see also [Hooker, 1988]). The general form for an implication is

$$A \to \Phi \tag{7.24}$$

where A is an action (transition firing) or the result of an action (place marking). Φ is a "well-formed" boolean expression describing the state of the system. It contains boolean variables (place markings) and any of the logical operators And, Or, Not, Implication, or Equivalence. Assuming the net is safe, (7.24) may be transformed into a set of linear inequalities through the following procedure.

First Φ must be written in conjunctive normal form (product-of-sums form). For example, the boolean expression $A + BC$ would be transformed to $(A + B)(A + C)$.

At this point, implication (7.24) has become

$$A \rightarrow \phi_1 \phi_2 \cdots \phi_g \tag{7.25}$$

where

$$\phi_i = \psi_1 + \psi_2 + \cdots + \psi_h \tag{7.26}$$

Each variable ψ is boolean and is valued *true* or *false* according to the marking of a corresponding Petri net place (which can only be 1 or 0 since the net is assumed to be safe).

The implication can now be broken down into a system of g simultaneous linear inequalities, where g is the number of product terms in (7.25). Each inequality will have the form

$$(1 - \mu_{\psi_1}) + (1 - \mu_{\psi_2}) + \cdots + (1 - \mu_{\psi_h}) + A \leq h \tag{7.27}$$

where h is number of summed terms in the corresponding ϕ value and μ_{ψ_i} is the marking of the place associated with ψ_i.

Example.   A system of inequalities will be derived to enforce the constraint

if transition 1 fires then places 1 and 2 must contain tokens

on a safe Petri net. The action, A, is the firing of transition 1, while Φ is the logical AND of the truth of places 1 and 2 containing tokens. Though it is a slight notational abuse, for convenience the implication is expressed as

$$q_1 \rightarrow \mu_1 \mu_2$$

which is in conjunctive normal form. Following the procedure outlined above, the transformation into a system of linear inequalities results in

$$\begin{aligned}
1 - \mu_1 + q_1 &\leq 1 \\
1 \quad \mu_2 \mid q_1 &\leq 1
\end{aligned}$$

or

$$\begin{aligned}
q_1 - \mu_1 &\leq 0 \\
q_1 - \mu_2 &\leq 0
\end{aligned}$$

which will insure that the original implication is true when the inequalities are enforced on the system behavior using either direct or indirect methods from section 7.2.

7.4   Constraints Involving Time

The control methods of this book are driven by sequences of events (firings of transitions) occurring in the plant. Some transitions fire before others, some may fire simultaneously. The states of the plant and controller evolve through time, but there is no direct representation of time. The controller may respond to the firing of transition t_i and then respond to the firing of transition t_j, but there is no indication of how much time has elapsed between the firings of these two transitions. This section will be used to discuss the issues that arise when time is introduced into the control framework.

The most common way of introducing time into a Petri net model of a system is through the use of *timed Petri nets* [Sifakis, 1979, Ramamoorthy and Ho, 1980]. A timed net works like an ordinary Petri net but includes a new function defined on either the transitions or the places of the net. The function indicates the amount of time required for particular transitions to fire or the amount of time that must elapse between the arrival of a token in a place and when it is allowed to take part in enabling and firing another transition. In many cases the function is simply a constant vector that indicates the timing requirements for each of the net's transitions or places, but it may also be quite complicated with the firing times relying on the state of the net, the current time, and other factors.

Timed nets are useful for a variety of temporal analysis tasks. They can be used to determine the mean variation or mean value of the Petri net state over time, to determine steady state values in the net's invariants, or to calculate cycle times in cyclic nets. A control designer may wish to use timed nets to optimize or otherwise regulate some of these criteria.

Normally timed nets are used when the designer wishes to improve the operation of the plant relative to some time based metric or simply wishes to analyze the temporal behavior of the plant. In these situations a hierarchical control structure [Antsaklis and Passino, 1993, Antsaklis and Kantor, 1995, Antsaklis et al., 1993a, Antsaklis et al., 1993b] can be used where optimization and analysis are performed after supervisory control has been used to insure certain safety constraints have been met. In these situations, the invariant based supervisory control technique may be seamlessly incorporated into the design. ·

Timed nets are a useful extension of ordinary nets because they do not alter the basic behavior of these nets, they simply provide more information about their evolution. This means that the standard PN definitions, including structural invariants, still hold true. A controller that enforces certain state constraints and sequential behaviors on an ordinary Petri net will enforce these same behaviors on the net after timing information is added to it.

The invariant based control method is implemented through new places and arcs that connect to existing plant transitions. If the timing information of the plant net were associated with the transitions, then the control method could be used without any changes in the method itself. Because the controller has no new transitions of its own, it is not necessary to establish any new timing properties for the controller. It will react to the firings of the plant and will evolve naturally using the plant's own timing. Control goals such as mutual exclusions, deadlock avoidance, regulation of finite resources, or avoiding forbidden states may be implemented on timed Petri nets using the method exactly as described.

Unfortunately, when timing periods are associated with transitions, the resulting behavior of the controlled plant may not be entirely what the control designer expected. Consider the constraint

$$\mu_1 + q_1 \leq 1$$

In an ordinary Petri net, transition firings are considered to be instantaneous. Because of this, the constraint above means two things: 1) place 1 may never have more than one token, and 2) transition 1 may not fire if p_1 contains a token. Now consider if this constraint were to be placed on a Petri net that contained timed transitions. The constraint would take on a third meaning: 3) place 1 may not contain any tokens while transition 1 is *in the process of firing.* In some cases, this may be exactly what the designer wants, however the designer must be aware of these subtle changes in the meaning of the constraint inequalities when designing the system.

It is possible to maintain the original meaning of the constraint inequalities by using nets that place their timing information on the places rather than the transitions. Transitions undergo instantaneous firing in these nets, just like in ordinary Petri nets. The rest of the discussion will deal with these kind of timed nets.

Because a period of time must elapse before a token can take part in the firing of a transition, timed nets require a variation in the implementation of the firing rules. One method involves keeping track of two separate marking vectors, μ_a and μ_w. The tokens represented by the vector μ_a are *available* to be used to enable transitions. The elements of μ_w are still in their *waiting* period and may not be used to enable transitions until this period ends. The transition enabling rule is changed to

$$D^- q \leq \mu_a$$

When a transition fires, tokens are transferred from μ_a to μ_w using

$$\mu_a \Leftarrow \mu_a - D^- q$$
$$\mu_w \Leftarrow \mu_w + D^+ q$$

The elements in μ_w then transfer to μ_a after the time requirement for the particular place they are in has elapsed. Let $\mu_{w_i}(q)$ be the number of tokens that were transferred to place i upon the firing q, i.e.,

$$\mu_{w_i}(q) = (D^+ q)_i$$

Let μ_{w_i} and μ_{a_i} be the total number of waiting and available tokens in place i. When the time period after the firing q for place i elapses,

$$\mu_{a_i} \Leftarrow \mu_{a_i} + \mu_{w_i}(q)$$
$$\mu_{w_i} \Leftarrow \mu_{w_i} - \mu_{w_i}(q)$$

These rules do not produce behavior that differs from that of an ordinary Petri net. A timed net can be transformed into an ordinary net by removing the information about the required time constraints on the places and combining the two timed marking vectors to form a single marking vector: $\mu = \mu_a + \mu_w$.

Constraints may be enforced on nets with timed places using the supervisory control techniques discussed in previous chapters. The controller will add places to the plant net, and because this is a timed net, timing information must be associated with these new controller places. The tokens in a controller place are used to keep track of the plant's state. They are the controller's bookkeeping device, and do not represent a process that requires lengthy amounts of time compared to the time associated with the evolution of the plant. For this reason, *the timing requirements associated with controller places are defined as zero.* The example below illustrates how standard constraints may be enforced on a time net.

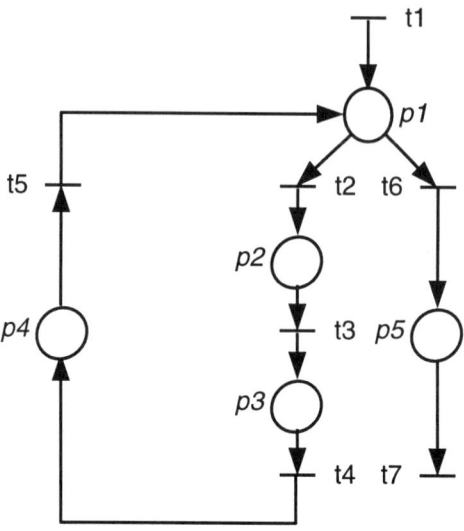

Figure 7.5.   A timed Petri net.

Example.   The Petri net of Figure 7.5 represents a buffered servicing system. Input arrives in the system through the firing of transition 1. Input may be directed through two paths. The left hand path performs several operations and then recycles the input back to the starting point in p_1. Input sent to the right hand path is serviced before it leaves the system. Tokens must spend a minimum amount of time in each place before they are permitted to be used to fire a transition.

$$\text{Times for } \begin{bmatrix} p_1 \\ p_2 \\ p_3 \\ p_4 \\ p_5 \end{bmatrix} = \begin{bmatrix} 1.0 \\ 0.5 \\ 3.0 \\ 2.5 \\ 5.0 \end{bmatrix} \text{ minutes}$$

There are two standard supervisory control constraints to be placed on the system. Place 2 is a buffer with a maximum load of 3:

$$\mu_2 \le 3 \tag{7.28}$$

Where $\mu_2 = \mu_{a2} + \mu_{w2}$. The second constraint is to prevent an illegal state/event combination:

$$\mu_3 + q_5 \le 1 \tag{7.29}$$

Neither of the two constraints directly involves time, so they can be implemented using the standard control synthesis procedures. The controlled system is shown in Figure 7.6. Constraint (7.29) is enforced using the "direct" rules of section 7.2.1. The timing for the control places is

$$\text{Times for } \begin{bmatrix} c_1 \\ c_2 \end{bmatrix} = \begin{bmatrix} 0 \\ 0 \end{bmatrix}$$

If the time period for the control places were greater than zero, then the controller might artificially delay the operation of the plant as it waited for its own tokens to be transferred from the "waiting" state to the "available" state.

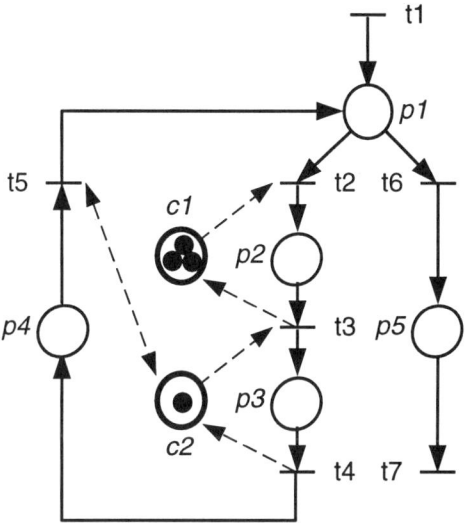

Figure 7.6.   The timed Petri net with a controller for two standard constraints.

Difficulties may arise when supervisory control specifications deal directly with time rather than events. Because the controller has no direct access to a clock, it is not possible to directly realize constraints that reference absolute time or relative timing offsets. Some of these kinds of situations may be tackled by including Petri net structures in the plant that the controller can use as an interface between its event-driven world and the world of continuous time.

Example. There are two more constraints to be placed on the plant of Figure 7.5. Unlike the two constraints enforced in the previous example, these constraints make reference to measured values of time:

1. Transition t_6 may not fire between 6:00 PM and 12:00 AM.

2. Transition t_7 may not fire until at least two minutes has elapsed since the last firing of transition t_4. If t_4 has never fired, then the firing of t_7 is unrestricted.

It is not possible to implement these constraints using only the structures from the previous example. The standard controller has no access to the current time of day, nor is it informed regarding the amount of time that separates the events to which it reacts. In order to enforce these constraints, new PN structures will be added to the plant, allowing an interface between the controller and time.

A one-way loop of timed PN places with a single token can be used to create a clock indicating the current time of day. For example, twenty-four places connected in a ring, with a one hour delay for each place, can be used to indicate the current hour of the day. It is not necessary that the delay in every place of the clock be equal, but *the firing rules for the clock must indicate that enabled transitions fire instantly.* Constraint 1 above indicates that we are concerned with the six hour period between 6:00 PM and 12:00 AM. The controller will gain access to the current time through the use of the two-place clock shown in the Figure 7.7a. The time requirements for the two places are

$$\text{Times for } \begin{bmatrix} p_6 \\ p_7 \end{bmatrix} = \begin{bmatrix} 18 \\ 6 \end{bmatrix} \text{ hours}$$

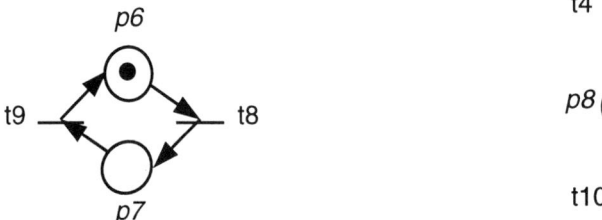

Figure 7.7a.   A two place time-of-day clock.

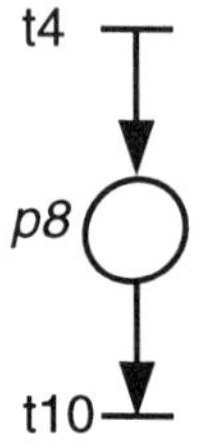

Figure 7.7b.   Place 8 contains tokens whenever fewer than two minutes has elapsed since the last firing of t_4.

The clock in Figure 7.7a is added to the net of Figure 7.6. It is initialized with a single token in p_6 at 12:00 midnight. After eighteen hours, the token in p_6 will be available to fire transition 8. The token in p_6 will transfer to p_7 at 6:00 PM, and it will remain there until midnight. Thus constraint 1 can be enforced by enforcing the inequality

$$\mu_7 + q_6 \leq 1 \tag{7.30}$$

on the revised plant.

Constraint 2 does not involve the time of day, rather it involves a relative offset in time after the firing of transition 4. Transition 7 must wait at least two minutes after the last firing of t_4 before it is allowed to fire. A timer is added to the plant to indicate when less than two minutes has elapsed after the last firing of t_4. This can be done using the net shown in Figure 7.7b, where the timing for for p_8 is two minutes. As with the clock of Figure 7.7a, the firing rules must indicate that t_{10} fires whenever it is enabled.

Using the net of Figure 7.7b, we know that transition 7 should never be allowed to fire when place 8 contains tokens. An initial guess might suggest that we then enforce the constraint $\mu_8 + q_7 \leq 1$, but constraint 2 says nothing about limiting the ability of t_4 to fire, and this constraint would prohibit p_8 from ever containing more than one token, indirectly inhibiting the freedom of t_4. Let M be the maximum number of tokens that could ever appear in place 8. This number could be determined through temporal of the plant, or the designer may simply choose M as a number so ridiculously large that t_4 could never fire that many times in a reasonable amount of time. Using this value, the following constraint is then placed on the plant,

$$\mu_8 + Mq_7 \leq M \tag{7.31}$$

This constraint will insure that t_7 never fires if place 8 contains any tokens. If M is too small, then transition t_4 may be indirectly inhibited by constraint (7.31), so it is necessary to determine a large enough value of M to avoid this situation.

Figure 7.8 shows the revised plant with added controller structures for constraints (7.30) and (7.31). As in the first example, the timing requirements for c_3 and c_4 are zero, as delays in these places would cause artificial delays in the plant.

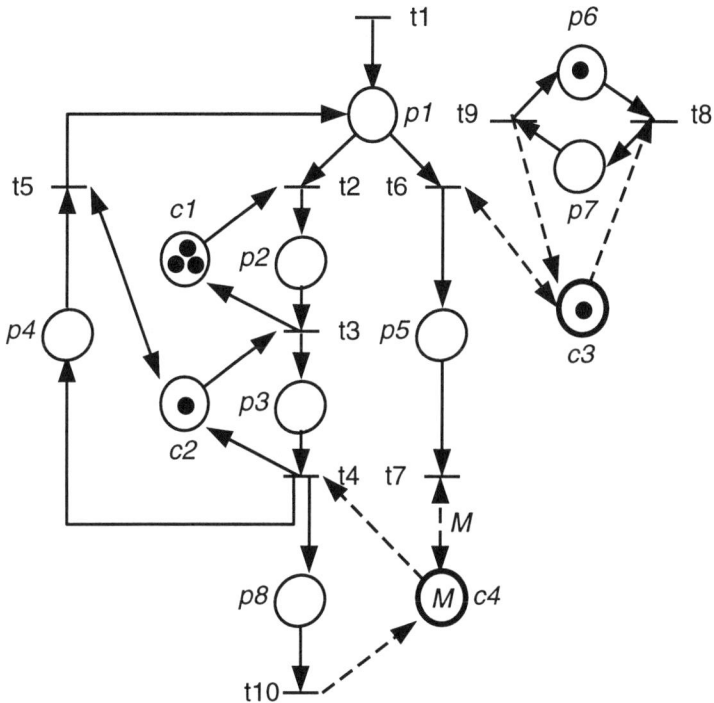

Figure 7.8.   The final plant model with supervisory controllers in place.

In summary, the control designer should be aware of the following points when introducing timing requirements into the invariant based control method.

- Defining time information on the places of a net, rather than its transitions, avoids certain ambiguities in the meanings of combined state/event constraints. Either method may be used, but the meanings of the constraints in each case must be understood.

- When time is defined on the places of a net, the timing requirements associated with the controls are defined as zero to avoid artificial delays in the evolution of the plant.

- Standard supervisory control constraints dealing with the plant state or mutual exclusions can be implemented seamlessly on place-timed Petri net plants.

- External structures with appropriately timed places may be added to the plant to provide an interface between the event sequences of supervisory control and real time.

- Real time clocks, such as that made by p_6 and p_7 in the example, may be added to the net in order to enforce constraints dealing with absolute time.

- Timers may be added to the plant, such as p_8 in the example, which are synchronized with the firing of plant transitions to enforce constraints dealing with relative time offsets.

- The accuracy of clocks and timers is insured by firing rules that insist on the instantaneous firing of enabled transitions for these structures.

- Upper bounds such as M in the example can be used to avoid unwanted consequences of state/event related constraints that deal with the conditions of timers and clocks.

- Timing constraints such as "transition 2 must fire within two minutes after any firing of transition 1" may be more difficult to enforce than the constraints covered in the examples above. Supervisory control techniques are not used, in general, to force any particular event in the evolution of the plant, rather they are used to restrict the plant's evolution from unwanted behavior. Timing requirements that demand that certain events take place would be more properly handled at another level of a hierarchical controller [Antsaklis and Passino, 1993, Antsaklis and Kantor, 1995, Antsaklis et al., 1993a, Antsaklis et al., 1993b].

7.5   Limitations of the Constraint Inequality

This chapter has demonstrated that many DES supervisory control problems can be solved by enforcing constraints of type (3.4) or (5.20) on a plant modeled by a Petri net, however these kind of constraints are not sufficient to realize all control goals. In this section, examples are used to illustrate certain problem areas for the proposed control method.

It is desirable to model DES's with Petri nets because the PN gives a simpler and more compact model of a process than an automaton with all states and transitions between states explicitly defined. However the compactness of the Petri net model can cause difficulties for certain arbitrary forbidden state problems as illustrated in the example below.

Example.   Consider the Petri net of Figure 7.9. We wish to prevent a single forbidden state $\mu_p = [1\ 1\ 2]^T$ (shown in the figure). This can not be enforced with maximal permissivity using constraints of the form $L\mu_p \le b$.

Suppose we tried to forbid the state $\mu = [1\ 1\ 2]^T$ with the following inequality:

$$\mu_1 - \mu_2 - \mu_3 \ne 0$$

If this constraint were implemented the controller would not be maximally permissive since it would disallow other states besides $[1\ 1\ 2]^T$, i.e., it would also disallow the states

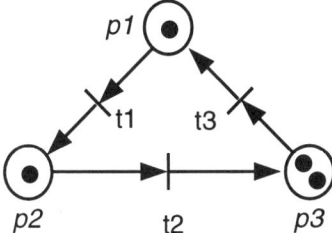

Figure 7.9.   The "forbidden state" of the example.

$[2\ 0\ 2]^T$ and $[2\ 2\ 0]^T$. A solution to this problem can be found if the net is bounded, which is true for this example. Observe that there are four tokens in the net, and the net can never have more than four tokens. We can write the constraint using base five, with the marking of each place being a digit in a base five number. Let the marking of place 1 be the one's place, place 2's marking be the fives place, and place 3's marking be the twenty-fives place. We can then write the constraint as

$$\mu_1 + 5\mu_2 + 25\mu_3 \neq 56 \tag{7.32}$$

since $(1 \times 5^0) + (1 \times 5^1) + (2 \times 5^2) = 56$.

A controller implementing equation (7.32) would be maximally permissive, however equation (7.32) can not be written in the form of equation (3.4). Equation (7.32) represents a *disjunction* of two inequalities

$$(\mu_1 + 5\mu_2 + 25\mu_3) \leq 55 \vee (\mu_1 + 5\mu_2 + 25\mu_3) \geq 57$$

Enforcing this constraint with invariant based controllers requires the extended PN controller form of section 5.4.

DES control is not only used to prevent forbidden states but also used to force the DES plant to realize a desired formal language [Ramadge and Wonham, 1989, Li and Wonham, 1993, Li and Wonham, 1994, Giua and DiCesare, 1994]. The following two examples illustrate language realization issues in light of the given control scheme.

Example.   The Petri net of Figure 7.10, which appears in [Giua and DiCesare, 1994], is unbounded: the number of tokens in p_2 has no upper limit. The transitions of the net are marked either a or b corresponding to the language symbols that are generated when a given transition fires. The language spoken by this net is

$$a^k b a^l b$$

where $k \geq l \geq 0$.

We wish to control the net to achieve the language

$$a^k b a^k b, \qquad k \geq 0 \tag{7.33}$$

which will lead from an initial state of $\mu = [1\ 0\ 0\ 0]^T$ to a desired final state of $\mu = [0\ 0\ 0\ 1]^T$. It would be possible to achieve language (7.33) by placing an *inhibitor arc* from p_2 to the rightmost b transition. The addition of inhibitor arcs prohibits the use of the simple vector and matrix descriptions of Petri nets and will not be considered as a

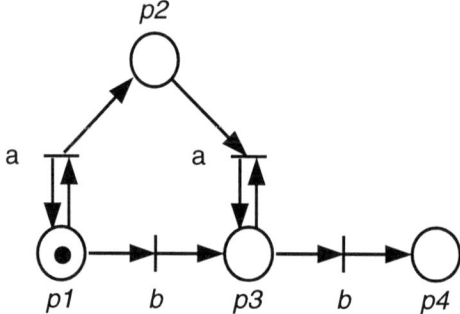

Figure 7.10.   An unbounded Petri net.

desirable solution here. Unfortunately a controller with the abilities of a Turing machine is required to solve the problem as stated, thus there is no (finite) Petri net controller that can force the net of Figure 7.10 to speak language (7.33).

Suppose we adjust the language in (7.33) to

$$a^k ba^k b, \qquad 0 \le k \le K \tag{7.34}$$

That is, we have put a bound on the number of times the first a transition will be allowed to fire. This problem can be solved with the place-invariant control method. The incidence matrix of the uncontrolled plant is

$$D_p = \begin{bmatrix} 0 & -1 & 0 & 0 \\ 1 & 0 & -1 & 0 \\ 0 & 1 & 0 & -1 \\ 0 & 0 & 0 & 1 \end{bmatrix}$$

The following constraint will enforce language (7.34):

$$\mu_2 + K\mu_4 \le K$$

Choosing $K = 3$ we obtain

$$\begin{aligned} D_c &= -[0\ 1\ 0\ 3]D_p \\ &= [-1\ 0\ 1\ -3] \end{aligned}$$

The controlled net is shown in Figure 7.11. Note that the value of K can be chosen arbitrarily high. The higher the value of K the closer the controlled language becomes to the actual desired language (7.33). In Figure 7.11, the transition between the controller place and the second b transition requires three (K) tokens in the controller place before the transition can be enabled.

Now suppose we change the language requirement to

$$a^K ba^K b \tag{7.35}$$

where K is a constant integer greater than zero. That is we want the plant to fire the a transitions a set number of times every time it runs. This can be accomplished by augmenting the controlled net of Figure 7.11 with a *memory*. The memory is used

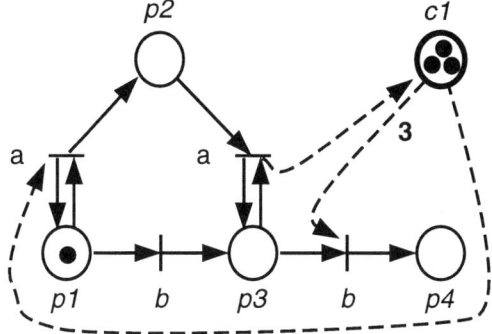

Figure 7.11.   The unbounded Petri net controlled.

to count the number of times the first a transition is fired and to not allow the first b transition to fire until all K a firings have occurred. The memory is implemented as a second controller place with a weight K arc connecting it to the first b transition. The controlled net is shown in Figure 7.12 with $K = 3$.

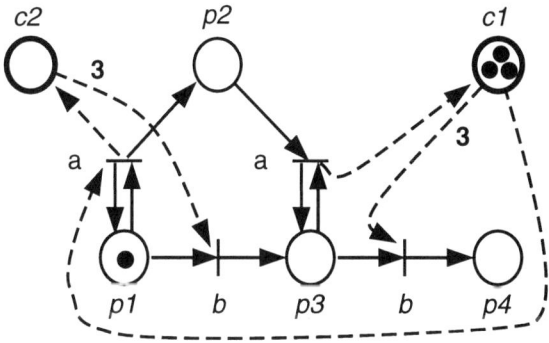

Figure 7.12.   The memory-augmented controlled net.

The controllers produced by the place invariant control method are dynamic, however the dynamics are used to simply keep track of the slack or excess tokens in a particular desired invariant. The fact that they are dynamic gives them a compact representation and allows them to be computed easily, but they have no more decision power than a (possibly infinite) static map of PN states to control-enabled transitions.

Proposition 7.6 Invariant based controllers enforce static state feedback control laws. The control law enforced by invariant based supervisors is a static map from the current plant state to disjoint sets of enabled and disabled transitions.

Proof. The proof of Theorem 3.2 indicates that the closed loop system of the plant and its invariant based controller contains the following set of place invariants.

$$L\mu_p + \mu_c = b$$

where μ_p is the state of the plant, μ_c is the state of the controller, and L and b are the constraint parameters. The sets of transitions that the controller will allow or disallow at any given point is determined by the controller state, and $\mu_c = b - L\mu_p$ is uniquely determined by the plant state.   ■

Language realization problems often require more than simple state feedback in order to be solved. Static controllers, mapping the current state to a set of allowed transitions, are usually not sufficient. Language realization requires the abilities that may be derived from static state feedback, but also requires some form of memory. In the previous example, the controller places were used to enforce an invariant as well as keep track of when a certain condition had occurred, however this problem could have been solved with static state feedback. In the following example the use of memory for language realization is further explored in a situation where static state feedback will not work.

Example. There are two minimal circuits in the Petri net of Figure 7.13.

circuit 1 : $p_1 \rightarrow p_2 \rightarrow p_3 \rightarrow p_1$
circuit 2 : $p_1 \rightarrow p_2 \rightarrow p_4 \rightarrow p_1$

We wish to control this net so that it alternates between the two paths, first choosing circuit 1, then 2, then 1, etc...

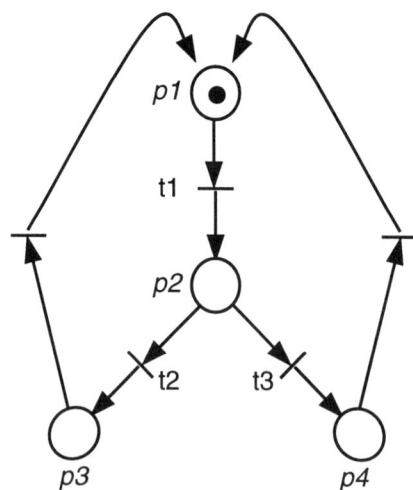

Figure 7.13.   The two-circuit net.

The incidence matrix of this Petri net is

$$D_p = \begin{bmatrix} -1 & 0 & 0 & 1 & 1 \\ 1 & -1 & -1 & 0 & 0 \\ 0 & 1 & 0 & -1 & 0 \\ 0 & 0 & 1 & 0 & -1 \end{bmatrix}$$

and its single place invariant is given by

$$X_p = \begin{bmatrix} 1 \\ 1 \\ 1 \\ 1 \end{bmatrix}$$

i.e., the entire net is an invariant.

A possible Petri net controller for this problem is given in Figure 7.14. The token in the controller jumps back and forth between the two controller places, alternately enabling transition 2 or transition 3.

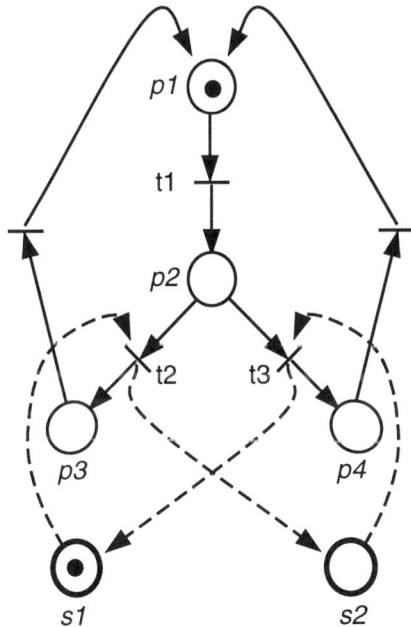

Figure 7.14.   The controlled two-path net.

The incidence matrix for the controller is

$$D_c = \begin{bmatrix} 0 & -1 & 1 & 0 & 0 \\ 0 & 1 & -1 & 0 & 0 \end{bmatrix}$$

and the invariants for the *controlled* net are given by

$$X^T \begin{bmatrix} D_p \\ D_c \end{bmatrix} = 0 \rightarrow X = \begin{bmatrix} 0 & 1 \\ 0 & 1 \\ 0 & 1 \\ 0 & 1 \\ 1 & 0 \\ 1 & 0 \end{bmatrix}$$

Note that *the controller places form their own invariant*, completely separate from the plant. Section 3.3 showed that, for an invariant based controller, each place in the controller forms an invariant with other places in the plant. Clearly the controller in Figure 7.14 could not be synthesized using this method. A method for using the invariant-based control method to obtain a solution like this is not obvious. The controller can be thought of as switching between two invariants involving $\{p_1, p_2, p_3\}$ and $\{p_1, p_2, p_4\}$, however a method for defining the constraint equations that would produce the controller shown in Figure 7.14 (or a similar, working controller) is not readily apparent.

8 EXAMPLE APPLICATIONS

Each of the sections of this chapter is intended to illustrate the synthesis techniques and procedures from the preceding chapters.

The "cat and mouse problem" of section 8.1.1 and the automated guided vehicle (AGV) coordination problem of section 8.2 illustrate the basic control design procedure when all of the transitions in the plant net are controllable and observable. The AGV example also includes an enforcement of "indirect" firing vector constraints (see section 7.2.2).

Plants with uncontrollable transitions are introduced for the cat and mouse in section 8.1.2 and for an "unreliable machine" in section 8.3. Both of these plants are also prone to deadlock. The deadlock avoidance methods of chapter 6 are used to prevent this condition and insure liveness.

The piston rod robotic assembly cell of section 8.4 illustrates the modeling of finite resources, reconfiguration due to sensor failures, and the characterization of admissible controls in the face of transition uncontrollability and unobservability.

Sections 8.5, 8.6, and 8.7 provide examples of handling constraints on events as well as states (section 7.2.1) using an asynchronous transfer mode switch, a three-tank process control system, and a hybrid (continuous/discrete) system.

8.1 The Cat and Mouse Problem

8.1.1 Supervision with all transitions controllable

The "cat and mouse" problem, introduced by [Wonham and Ramadge, 1987], is a popular example in the field of discrete event system control. The problem involves a maze of five rooms where a cat and a mouse can circulate. The rooms are connected with doors through which the animals can pass as shown in Figure 8.1.

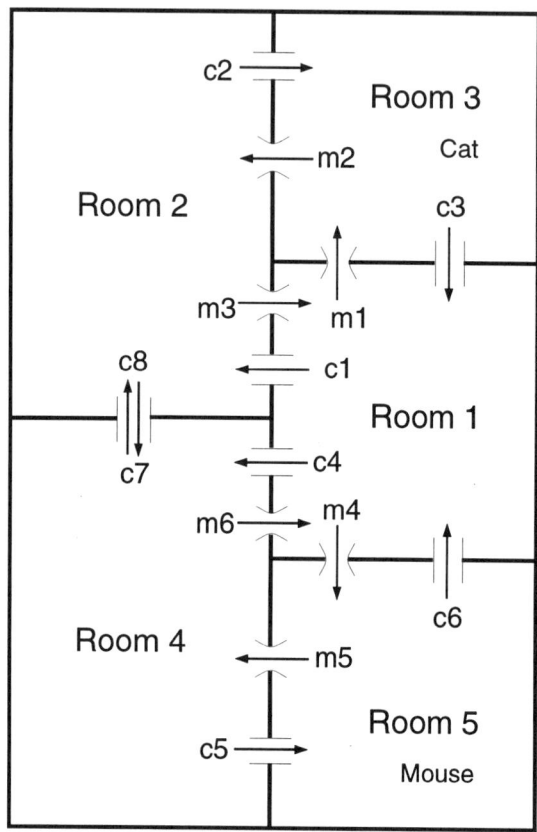

Figure 8.1.   The cat and mouse maze.

The problem is to control the doors so that the cat and the mouse can never be in the same room at the same time. The controller should be maximally permissive: it should grant maximum freedom of movement to both the cat and the mouse. The Petri net model of the cat and mouse problem is taken from [Boissel, 1993] and is shown in Figure 8.2. The upper net concerns the cat while the lower net concerns the mouse. Each net has five places that model the five rooms of the maze. The transitions model the ability of each animal to pass from one room to the other according to Figure 8.1.

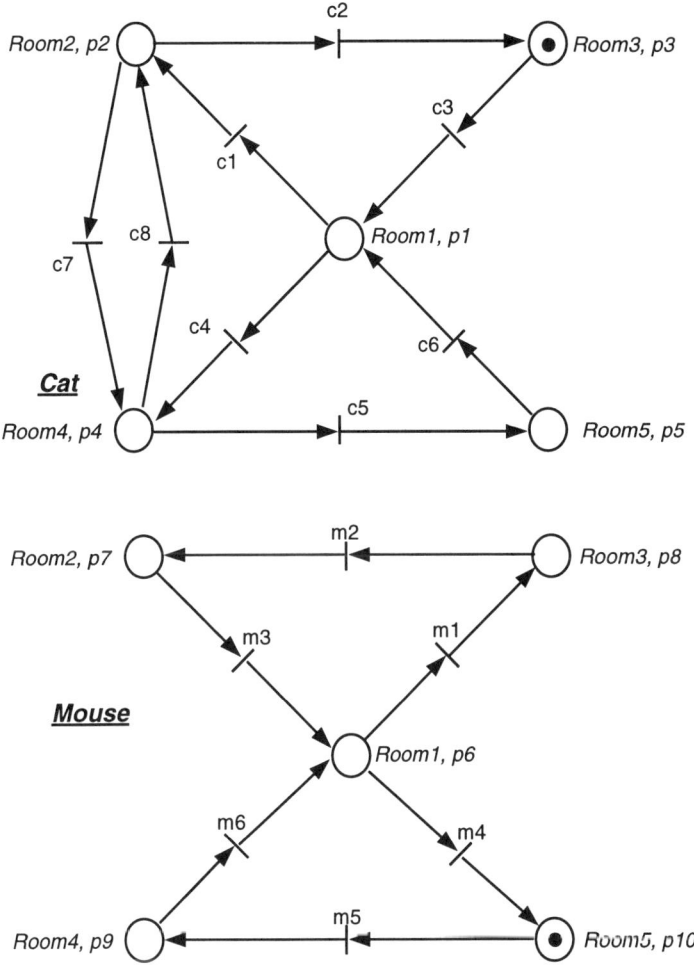

Figure 8.2.   The Petri net model of the cat and mouse problem.

The incidence matrix of the Petri net model is

$$
D_p =
\begin{bmatrix}
-1 & 0 & 1 & -1 & 0 & 1 & 0 & 0 & 0 & 0 & 0 & 0 & 0 & 0 \\
1 & -1 & 0 & 0 & 0 & 0 & -1 & 1 & 0 & 0 & 0 & 0 & 0 & 0 \\
0 & 1 & -1 & 0 & 0 & 0 & 0 & 0 & 0 & 0 & 0 & 0 & 0 & 0 \\
0 & 0 & 0 & 1 & -1 & 0 & 1 & -1 & 0 & 0 & 0 & 0 & 0 & 0 \\
0 & 0 & 0 & 0 & 1 & -1 & 0 & 0 & 0 & 0 & 0 & 0 & 0 & 0 \\
0 & 0 & 0 & 0 & 0 & 0 & 0 & 0 & -1 & 0 & 1 & -1 & 0 & 1 \\
0 & 0 & 0 & 0 & 0 & 0 & 0 & 0 & 0 & 1 & -1 & 0 & 0 & 0 \\
0 & 0 & 0 & 0 & 0 & 0 & 0 & 0 & 1 & -1 & 0 & 0 & 0 & 0 \\
0 & 0 & 0 & 0 & 0 & 0 & 0 & 0 & 0 & 0 & 0 & 0 & 1 & -1 \\
0 & 0 & 0 & 0 & 0 & 0 & 0 & 0 & 0 & 0 & 0 & 1 & -1 & 0
\end{bmatrix}
$$

The presence of a token in a place indicates that the animal modeled by the token is in the room modeled by the particular place. Initially the cat is in room 3 and the mouse is in room 5, so the initial marking is

$$\mu_{p_0} = \begin{bmatrix} \mu_1 & \mu_2 & \cdots & \mu_{10} \end{bmatrix}^T = \begin{bmatrix} 0 & 0 & 1 & 0 & 0 & 0 & 0 & 0 & 0 & 1 \end{bmatrix}^T$$

Assume that all the doors of the maze are controllable. The control goal is to ensure that the cat and the mouse are never in the same room simultaneously. This means that each pair of places, one from the upper net and one from the lower net, that model the same room must never contain more than one token. This requirement is translated into the following five constraints, which are forced into equalities by introducing slack variables.

$$
\begin{aligned}
\mu_1 + \mu_6 &\leq 1 &\Rightarrow& \quad \mu_1 + \mu_6 + \mu_{c_1} = 1 \\
\mu_2 + \mu_7 &\leq 1 &\Rightarrow& \quad \mu_2 + \mu_7 + \mu_{c_2} = 1 \\
\mu_3 + \mu_8 &\leq 1 &\Rightarrow& \quad \mu_3 + \mu_8 + \mu_{c_3} = 1 \\
\mu_4 + \mu_9 &\leq 1 &\Rightarrow& \quad \mu_4 + \mu_9 + \mu_{c_4} = 1 \\
\mu_5 + \mu_{10} &\leq 1 &\Rightarrow& \quad \mu_5 + \mu_{10} + \mu_{c_5} = 1
\end{aligned}
\tag{8.1}
$$

The five slack variables correspond to five controller places, i.e.,

$$\mu_c = \begin{bmatrix} \mu_{c_1} & \mu_{c_2} & \mu_{c_3} & \mu_{c_4} & \mu_{c_5} \end{bmatrix}^T$$

Using the matrix notation of equation (3.5) we have

$$L = \begin{bmatrix} 1 & 0 & 0 & 0 & 0 & 1 & 0 & 0 & 0 & 0 \\ 0 & 1 & 0 & 0 & 0 & 0 & 1 & 0 & 0 & 0 \\ 0 & 0 & 1 & 0 & 0 & 0 & 0 & 1 & 0 & 0 \\ 0 & 0 & 0 & 1 & 0 & 0 & 0 & 0 & 1 & 0 \\ 0 & 0 & 0 & 0 & 1 & 0 & 0 & 0 & 0 & 1 \end{bmatrix} \qquad b = \begin{bmatrix} 1 \\ 1 \\ 1 \\ 1 \\ 1 \end{bmatrix}$$

The controller incidence matrix is given by equation (3.10)

$$
D_c = -L D_p =
\begin{bmatrix}
1 & 0 & -1 & 1 & 0 & -1 & 0 & 0 & 1 & 0 & -1 & 1 & 0 & -1 \\
-1 & 1 & 0 & 0 & 0 & 0 & 1 & -1 & 0 & -1 & 1 & 0 & 0 & 0 \\
0 & -1 & 1 & 0 & 0 & 0 & 0 & -1 & 1 & 0 & 0 & 0 & 0 & 0 \\
0 & 0 & 0 & -1 & 1 & 0 & -1 & 1 & 0 & 0 & 0 & 0 & -1 & 1 \\
0 & 0 & 0 & 0 & -1 & 1 & 0 & 0 & 0 & 0 & 0 & -1 & 1 & 0
\end{bmatrix}
\tag{8.2}
$$

The initial marking of the slack places is given by equation (3.11)

$$\mu_{c_0} = 1 - L\mu_{p_0} = \begin{bmatrix} 1 & 1 & 0 & 1 & 0 \end{bmatrix}^T$$

The supervisor prevents the cat and mouse from ever entering the same room but allows all other legal moves. The controlled cat and mouse system is shown in Figure 8.3. Dashed arcs and thick-lined places are used to represent the elements of the controller Petri net.

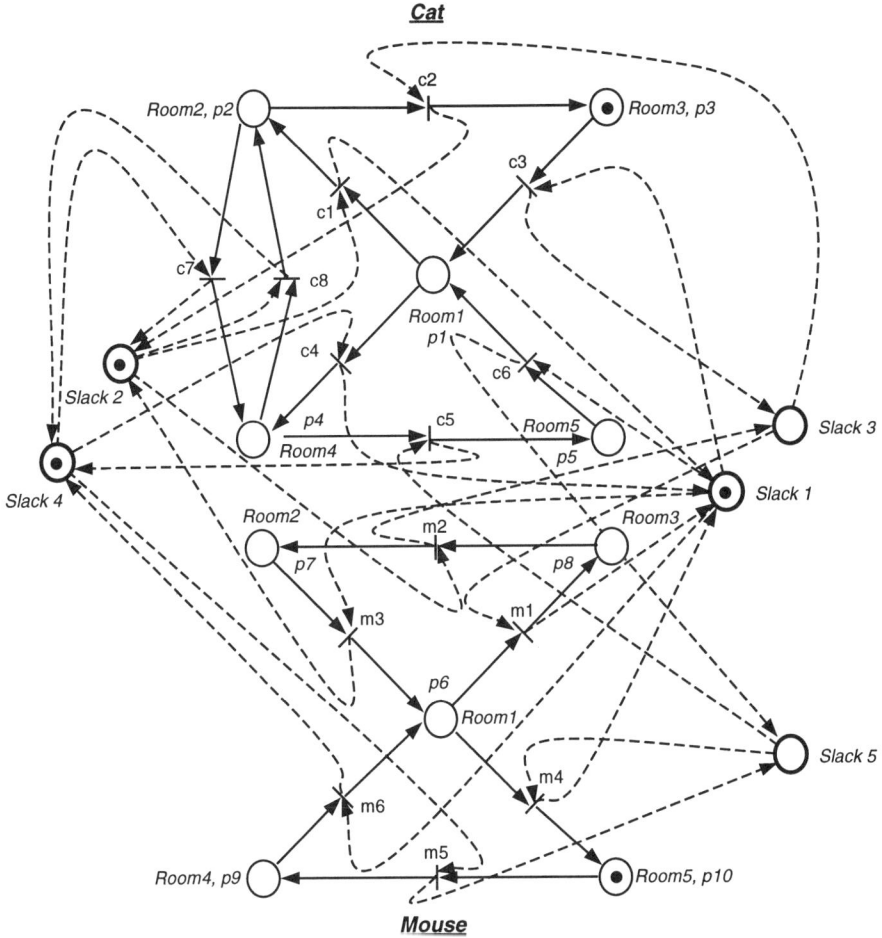

Figure 8.3.   The cat and mouse Petri net and its Petri net controller.

8.1.2   Introduction of uncontrollable transitions

The cat and mouse problem presented in [Wonham and Ramadge, 1987] specifies that transitions c_7 and c_8 are uncontrollable. Following the techniques of chapter 5 the D_{uc} matrix is then constructed from the seventh and eighth columns of D_p. Admissible constraints will satisfy $LD_{uc} \leq 0$.

$$LD_{uc} = \begin{bmatrix} 0 & 0 \\ -1 & 1 \\ 0 & 0 \\ 1 & -1 \\ 0 & 0 \end{bmatrix}$$

The second and fourth constraints yield positive numbers in the product LD_{uc}, so they must be transformed.

An addition of the fourth row of D_{uc} to the second row LD_{uc} will eliminate the positive number of that row. This indicates the constraint transformation

$$\mu_2 + \mu_7 \leq 1 \Rightarrow \mu_2 + \mu_4 + \mu_7 \leq 1$$

Similarly, an addition of the second row of D_{uc} to the fourth row LD_{uc} will eliminate the positive number of that row.

$$\mu_4 + \mu_9 \leq 1 \Rightarrow \mu_2 + \mu_4 + \mu_9 \leq 1$$

This leaves us with the following following set of admissible constraints

$$
\begin{aligned}
\mu_1 + \mu_6 &\leq 1 \\
\mu_2 + \mu_4 + \mu_7 &\leq 1 \\
\mu_3 + \mu_8 &\leq 1 \\
\mu_2 + \mu_4 + \mu_9 &\leq 1 \\
\mu_5 + \mu_{10} &\leq 1
\end{aligned}
$$

The second and fourth constraints both contain the expression $\mu_2 + \mu_4$ added to the marking of a place in the mouse's subnet. Because this subnet (places m_1 through m_5) forms a place invariant, we know that

$$\mu_7 + \mu_9 \leq 1$$

is always true due to the structure of the plant. Thus we can reduce the two constraints involving $\mu_2 + \mu_4$ into a single constraint

$$
\left.
\begin{aligned}
\mu_2 + \mu_4 + \mu_7 &\leq 1 \\
\mu_2 + \mu_4 + \mu_9 &\leq 1
\end{aligned}
\right\} \Rightarrow \mu_2 + \mu_4 + \mu_7 + \mu_9 \leq 1
$$

without any loss of generality or permissivity.

An analysis of the plant under the current set of four admissible constraints reveals that there is the potential for deadlock. Suppose that the mouse were in room 2 and the cat were in room 1. The mouse can only move to room 1, so it is blocked by the supervisor. The cat can move to room 2 or 4, but it is also blocked by the supervisor because of the uncontrollable connection the cat has between rooms 2 and 4.

Fortunately, as indicated in chapter 6, this potential deadlock can be eliminated by the imposition of a further linear inequality causing the occupation of rooms 1, 2, and 4 to be mutually exclusive to either the cat or the mouse. This inequality is achieved by combining the constraint $\mu_1 + \mu_6 \leq 1$ with the previously constructed constraint $\mu_2 + \mu_4 + \mu_7 + \mu_9 \leq 1$. There are now three different inequalities that will be enforced on the marking behavior of the plant in order to insure both the safety of the mouse and eternal freedom of movement for both animals.

$$
\begin{aligned}
\mu_1 + \mu_2 + \mu_4 + \mu_6 + \mu_7 + \mu_9 &\leq 1 \\
\mu_3 + \mu_8 &\leq 1 \\
\mu_5 + \mu_{10} &\leq 1
\end{aligned}
\tag{8.3}
$$

At this point the controller can be computed using the method of chapter 3. The controlled plant is shown in Figure 8.4.

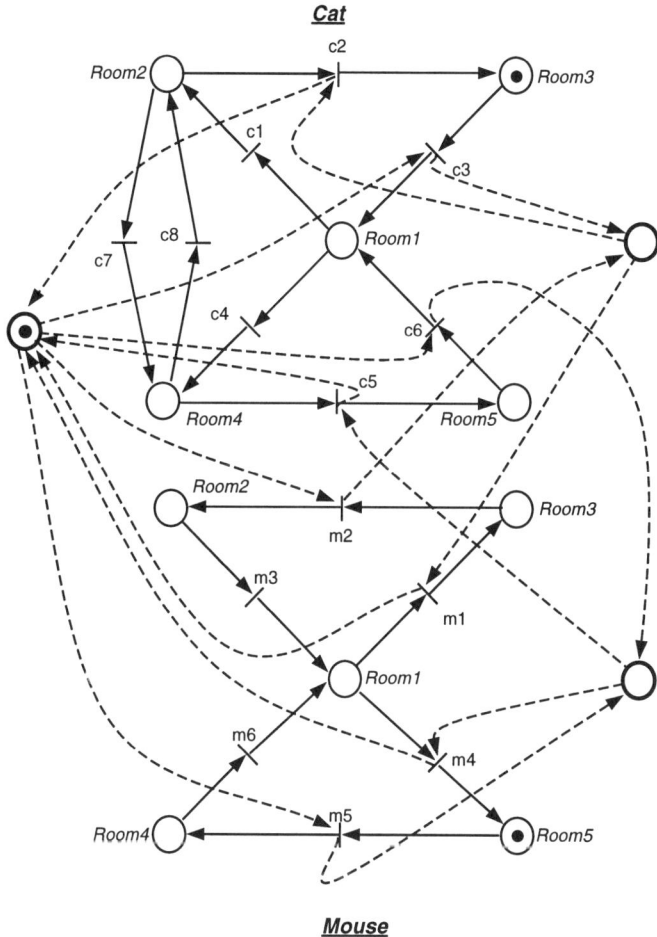

Figure 8.4. The cat and mouse Petri net and controller with transitions c_7 and c_8 uncontrollable.

8.2 Automated Guided Vehicle Coordination

This example, concerning a flexible manufacturing cell, was introduced by [Holloway and Krogh, 1990]. The cell, shown in Figure 8.5, includes three workstations, two part-receiving stations and one completed parts station. There are five automated guided vehicles (AGV's) that transport material between the stations. The vehicles and parts are modeled by tokens in the net (see [Holloway and Krogh, 1990]). The vehicular routes intersect on the plant floor, and consequently, there are zones, represented by the shaded regions in Figure 8.5, in which two vehicles could be present at the same time. This situation will be forbidden by the supervisor. An additional constraint is placed on the system, regarding the pickup of parts from the Input Parts Station, that illustrates the indirect method of firing vector constraint enforcement from section 7.2.2.

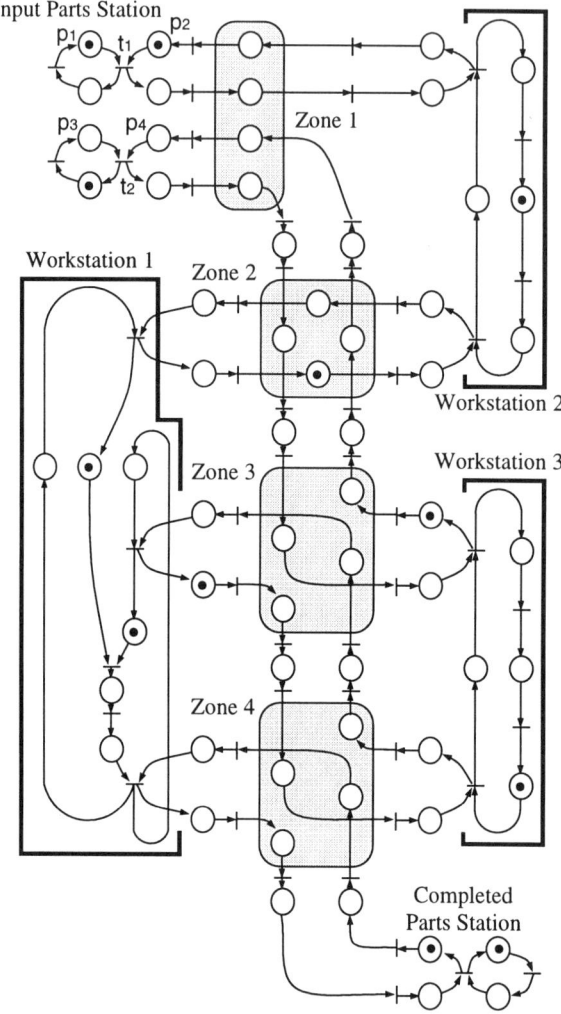

Figure 8.5.   The Automated Guided Vehicle Petri Net.

The constraints concerning the presence of vehicles in the intersection zones are expressed by the following inequalities

$$\sum_{i \in Z_1} \mu_i \ \leq \ 1$$

$$\sum_{i \in Z_2} \mu_i \leq 1$$

$$\sum_{i \in Z_3} \mu_i \leq 1$$

$$\sum_{i \in Z_4} \mu_i \leq 1$$

where Z_j is the set of indices of places which make up zone j. Four slack variables $(\mu_{c_1}, \ldots, \mu_{c_4})$ are introduced to make the inequalities become equalities

An additional specification is to be enforced concerning the Input Parts Station. There is only room for one AGV to enter the station at any time, so we will prevent the occurrence of two AGV's attempting to simultaneously pick up parts. Writing this constraint as an inequality, we have

$$q_1 + q_2 \leq 1 \tag{8.4}$$

where t_1 and t_2 are the transitions indicating that a part has been removed by one of the two AGV's that service the station. Note that the actual purpose of this constraint is not to prevent the simultaneous pick up of parts, but to prevent the two AGV's from being in the input station at the same time. *We are interested in preventing states that would enable firings such that $q_1 + q_2 > 1$, rather than preventing this firing itself.* This is an example of a firing vector constraint that should be handled by the "indirect" rules of section 7.2.2.

The first step of the procedure involves introducing dummy places into the net to represent the firing of transitions 1 and 2 as illustrated in Figure 7.1. Constraint (8.4) is then rewritten

$$\mu_1' + \mu_2' \leq 1 \tag{8.5}$$

and, as described in section 7.2.2, *transitions 1 and 2 are now treated as if they were uncontrollable, which makes (8.5) an inadmissible constraint.* It will need to be transformed using the techniques of chapter 5 into a constraint that prevents the states that could lead to (8.5) being violated, which is the desired goal of the indirect firing vector constraint.

Applying techniques from chapter 5, (8.5) is transformed into the following admissible constraint.

$$\mu_1 + \mu_2 + \mu_3 + \mu_4 + 2(\mu_1' + \mu_2') \leq 2(1 + 1) - 1$$

That is, $R_1 = \begin{bmatrix} 1 & 1 & 1 & 1 & 0 & 0 & \cdots \end{bmatrix}$ and $R_2 = 2$, which meet assumptions (4.21) and (4.22) of Lemma 4.10.

Now that an admissible constraint has been found, the dummy places μ_1' and μ_2' are eliminated and the plant is collapsed back into its original form. The constraint now becomes

$$\mu_1 + \mu_2 + \mu_3 + \mu_4 \leq 3$$

An additional slack variable makes the transformed constraint an equality and introduces a fifth controller place, μ_{c_5}. D_c and μ_{c_0} are computed using equations (3.10) and (3.11). The supervised plant is shown in Figure 8.6.

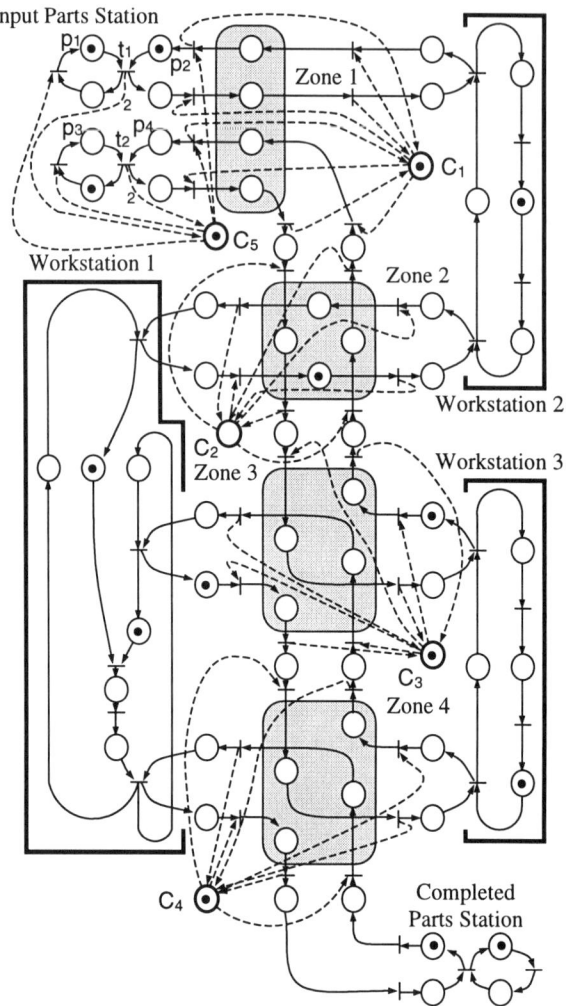

Figure 8.6.   The Controlled AGV Net.

8.3   The Unreliable Machine

8.3.1   Plant Model

The plant of Figure 8.7 features an "unreliable machine" (see [Desrochers and Al-Jaar, 1995] and [Moody et al., 1995b]). The machine is used to process parts from an input queue, completed parts are moved to an output queue by an automated guided vehicle (AGV). The machine is considered unreliable because it is possible that it may break down and damage a part during operation. This behavior is captured in the plant model. Damaged parts are moved to a separate queue by a second AGV. The Petri net model of the plant is shown in Figure 8.8, and a description of the various places and transitions is given in table 8.1.

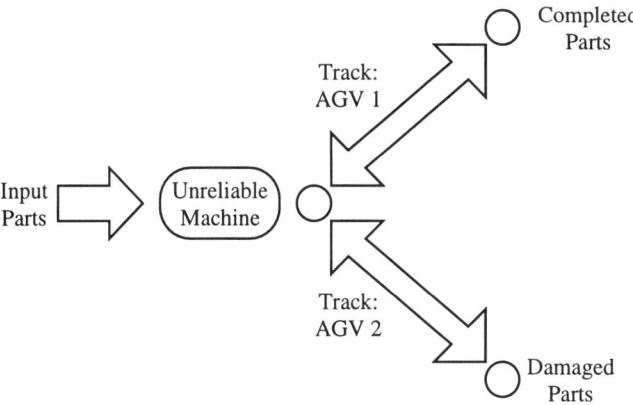

Figure 8.7.   Basic operation of the plant.

The plant model has two uncontrollable transitions, t_2 and t_6. Transition t_6 represents machine break down and so obviously can not be controlled. Transition t_2 is considered uncontrollable because the controller can not force the machine to instantly finish a part that is not yet completed, nor does it direct the machine to stop working on an unfinished part.

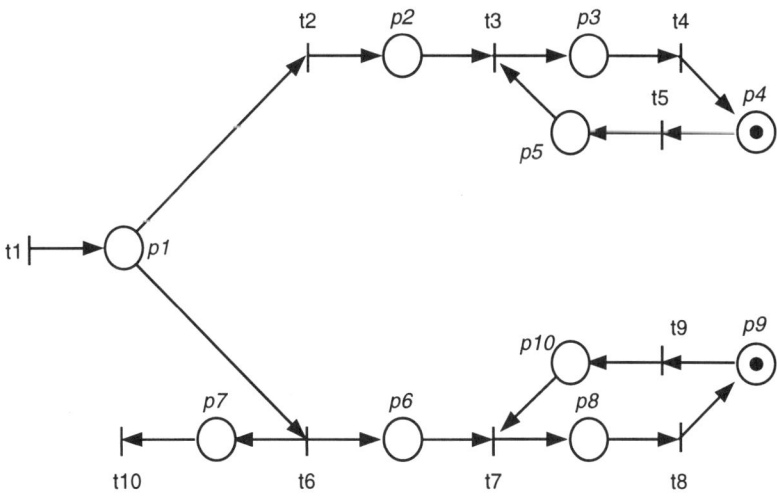

Figure 8.8.   Petri net model of the unreliable machine plant.

Table 8.1.   Place and transition descriptions for Figure 8.8.

	Places
p_1	Machine is "up and busy," part is being processed.
p_2	Part is waiting for transfer to completed-parts queue.
p_3	Part is being carried to completed-parts queue by AGV 1.
p_4	AGV 1 is free, away from part pick-up position.
p_5	AGV 1 is at pick-up position at machine.
p_6	Part is waiting for transfer to damaged-parts queue.
p_7	Machine is waiting to be repaired.
p_8	Part is being carried to damaged-parts queue by AGV 2.
p_9	AGV 2 is free, away from part pick-up position.
p_{10}	AGV 2 is at pick-up position at machine.

	Transitions
t_1	Part moves from input queue to machine.
t_2	*Uncontrollable:* Part processing is complete.
t_3	Part is picked up by AGV 1.
t_4	Part is deposited in completed-parts queue by AGV 1.
t_5	AGV 1 moves into pick-up position at machine.
t_6	*Uncontrollable:* Machine fails, part is damaged
t_7	Part is picked up by AGV 2.
t_8	Part is deposited in damaged-parts queue by AGV 2.
t_9	AGV 2 moves into pick-up position at machine.
t_{10}	Machine is repaired.

The plant has the following incidence matrix.

$$
D_p =
\begin{bmatrix}
1 & -1 & 0 & 0 & 0 & -1 & 0 & 0 & 0 & 0 \\
0 & 1 & -1 & 0 & 0 & 0 & 0 & 0 & 0 & 0 \\
0 & 0 & 1 & -1 & 0 & 0 & 0 & 0 & 0 & 0 \\
0 & 0 & 0 & 1 & -1 & 0 & 0 & 0 & 0 & 0 \\
0 & 0 & -1 & 0 & 1 & 0 & 0 & 0 & 0 & 0 \\
0 & 0 & 0 & 0 & 0 & 1 & -1 & 0 & 0 & 0 \\
0 & 0 & 0 & 0 & 0 & 1 & 0 & 0 & 0 & -1 \\
0 & 0 & 0 & 0 & 0 & 0 & 1 & -1 & 0 & 0 \\
0 & 0 & 0 & 0 & 0 & 0 & 0 & 1 & -1 & 0 \\
0 & 0 & 0 & 0 & 0 & 0 & -1 & 0 & 1 & 0 \\
0 & 0 & 1 & 0 & -1 & 0 & 1 & 0 & -1 & 0 \\
-1 & 0 & 1 & 0 & 0 & 0 & 1 & 0 & 0 & 0 \\
-1 & 1 & 0 & 0 & 0 & 0 & 0 & 0 & 0 & 1
\end{bmatrix}
\tag{8.6}
$$

The incidence matrix of the uncontrollable portion of the plant, D_{uc}, consists of the second and sixth columns of D_p.

8.3.2   Resource and Safety Constraints

There is room for only one AGV at a time to approach the machine and pick up a part, be it completed or damaged. Access to the machine is then restricted with the constraint

$$\mu_5 + \mu_{10} \leq 1 \tag{8.7}$$

There is also only room for one part, once the machine is done with it, at the AGV pickup position. Thus places p_2 and p_6 must remain exclusive:

$$\mu_2 + \mu_6 \leq 1 \tag{8.8}$$

Finally, if the machine breaks down, we must wait until the machine is repaired before starting work on a new part. This constraint corresponds to the exclusivity of places p_1 and p_7:

$$\mu_1 + \mu_7 \leq 1 \tag{8.9}$$

Constraints (8.7) through (8.9) are now represented in matrix form

$$\underbrace{\begin{bmatrix} 0 & 0 & 0 & 0 & 1 & 0 & 0 & 0 & 0 & 1 \\ 0 & 1 & 0 & 0 & 0 & 1 & 0 & 0 & 0 & 0 \\ 1 & 0 & 0 & 0 & 0 & 0 & 1 & 0 & 0 & 0 \end{bmatrix}}_{L} \mu_p \leq \underbrace{\begin{bmatrix} 1 \\ 1 \\ 1 \end{bmatrix}}_{b}$$

The condition of Corollary 4.7 is checked to determine if the constraints are admissible with respect to the plant's uncontrollable transitions.

$$LD_{uc} = \begin{bmatrix} 0 & 0 \\ 1 & 1 \\ -1 & 0 \end{bmatrix}$$

In order to meet the condition of Corollary 4.7 we need $LD_{uc} \leq 0$. Row two of the matrix fails this test.

Row operations on $\begin{bmatrix} D_{uc} \\ LD_{uc} \end{bmatrix}$ are now performed to obtain a transformed constraint matrix that is admissible.

$$
\begin{bmatrix}
-1 & -1 \\
1 & 0 \\
0 & 0 \\
0 & 0 \\
0 & 0 \\
0 & 1 \\
0 & 1 \\
0 & 0 \\
0 & 0 \\
0 & 0
\end{bmatrix}
\begin{bmatrix}
0 & 0 \\
1 & 1 \\
-1 & 0
\end{bmatrix}
\quad
\begin{array}{c}
\Rightarrow \\
\text{Row } 12 = \text{Row } 1 + \text{Row } 12
\end{array}
\quad
\begin{bmatrix}
-1 & -1 \\
1 & 0 \\
0 & 0 \\
0 & 0 \\
0 & 0 \\
0 & 1 \\
0 & 1 \\
0 & 0 \\
0 & 0 \\
0 & 0
\end{bmatrix}
\begin{bmatrix}
0 & 0 \\
0 & 0 \\
-1 & 0
\end{bmatrix}
$$

The LD_{uc} portion of the matrix now has no positive elements. The row operation above corresponds to the addition of μ_1 to constraint (8.8), i.e., $\mu_2 + \mu_6 \leq 1$ becomes $\mu_1 + \mu_2 + \mu_6 \leq 1$. Equivalently, using the R_1 and R_2 matrices of section 4.5, we have

$$
R_1 = \begin{bmatrix}
0 & 0 & 0 & 0 & 0 & 0 & 0 & 0 & 0 & 0 \\
1 & 0 & 0 & 0 & 0 & 0 & 0 & 0 & 0 & 0 \\
0 & 0 & 0 & 0 & 0 & 0 & 0 & 0 & 0 & 0
\end{bmatrix}
\qquad R_2 = I
$$

Given an admissible set of constraints, the controller is now calculated

$$
D_c = -L'D_p = -(R_1 + R_2 L)D_p =
$$
$$
\begin{bmatrix}
0 & 0 & 1 & 0 & -1 & 0 & 1 & 0 & -1 & 0 \\
-1 & 0 & 1 & 0 & 0 & 0 & 1 & 0 & 0 & 0 \\
-1 & 1 & 0 & 0 & 0 & 0 & 0 & 0 & 0 & 1
\end{bmatrix}
$$

$$
\mu_{c_0} = b' - L'\mu_{p_0} = R_2(b+1) - 1 - (R_1 + R_2 L)\mu_{p_0} = \begin{bmatrix} 1 \\ 1 \\ 1 \end{bmatrix}
$$

Figure 8.9 shows the supervisor added to the plant Petri net.

8.3.3   Deadlock Avoidance

It will be shown below that the combined action of the supervisory mechanisms above create the potential for the plant to become deadlocked. There are seven minimal siphons (see section 2.3) in the supervised model of the plant. Five of these, $\{p_1, p_7, c_3\}$, $\{p_1, p_2, p_6, c_2\}$, $\{p_3, p_4, p_5\}$, $\{p_5, p_{10}, c_1\}$ and $\{p_8, p_9, p_{10}\}$ also correspond to place invariants. Each of these invariants is marked, and thus these siphons can be considered either trap or invariant-controlled (see section 6.2.1). However, the remaining two siphons are uncontrolled:

$$
\begin{aligned}
S_1 &= \{p_1, p_2, p_{10}, c_1, c_2\} \\
S_2 &= \{p_1, p_5, p_6, c_1, c_2\}
\end{aligned}
$$

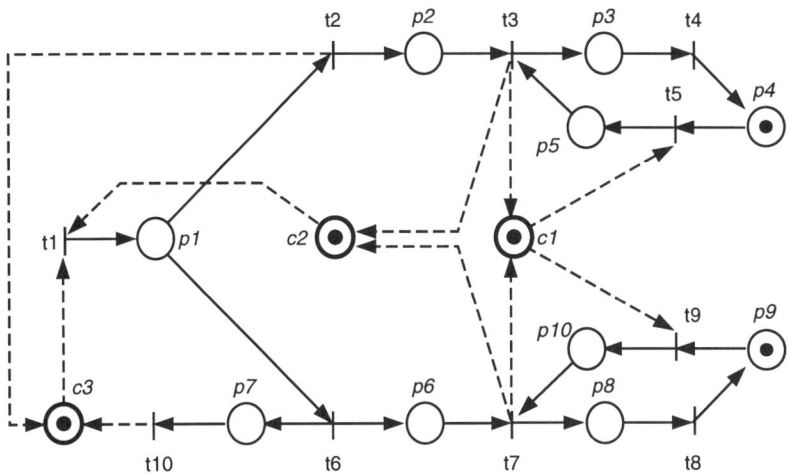

Figure 8.9.   The supervised unreliable machine, before accounting for deadlock avoidance.

Note that both uncontrolled siphons involve places c_1 and c_2. The interaction of the mutual exclusions being enforced by these two places can result in a deadlock condition. Suppose that AGV 1 moves into position at the machine while it is working on a part, i.e., t_5 fires while $\mu_1 = 1$. Now suppose the machine breaks down (t_6 fires). AGV 1 is stuck waiting for a completed part, and AGV 2 is stuck waiting for AGV 1 to move out of the way: deadlock has occurred. This deadlock condition corresponds to S_1, S_2 represents the analogous situation with the roles of the two machines reversed.

According to Proposition 6.3 and the method of section 6.3.2, we will attempt to avoid deadlock by insuring that these two siphons remain marked. Further supervisors will be added to cause the two siphons to become "controlled," i.e., for each siphon, a control place will be created that insures that sum of the tokens in the siphon remains greater than or equal to one. Using the notation $L\mu_p \leq b$ and assignment (6.9), we have,

$$L = \begin{bmatrix} -1 & -1 & 0 & 0 & 0 & 0 & 0 & 0 & 0 & -1 & -1 & -1 & 0 \\ -1 & 0 & 0 & 0 & -1 & -1 & 0 & 0 & 0 & 0 & -1 & -1 & 0 \end{bmatrix}$$

$$b = \begin{bmatrix} -1 \\ -1 \end{bmatrix}$$

where rows 1 and 2 correspond to the conditions for S_1 and S_2 respectively. Note that c_1 through c_3 are now being treated as places p_{11} through p_{13} in the plant.

Just as in the previous section, the condition of Corollary 4.7 is checked to determine if the constraint matrix is admissible.

$$LD_{uc} = \begin{bmatrix} 0 & 1 \\ 1 & 0 \end{bmatrix}$$

We need all elements of LD_{uc} to be nonpositive to meet inequality (4.19). If the supervisor were created using the given value of L, the first row of LD_{uc} indicates that it would attempt to achieve its goal by inhibiting transition t_6, which corresponds to machine break down. Unfortunately the unreliable machine is not impressed by requests from the controller to simply not break. A similar situation holds with row 2 of LD_{uc} and t_2. We will transform the constraint to eliminate the influence of the controller on these two uncontrollable transitions.

Again, the transformed constraint is constructed using positive row operations on $\begin{bmatrix} D_{uc} \\ LD_{uc} \end{bmatrix}$ to eliminate the positive numbers in LD_{uc}.

$$
\begin{bmatrix}
-1 & -1 \\
1 & 0 \\
0 & 0 \\
0 & 0 \\
0 & 0 \\
0 & 1 \\
0 & 1 \\
0 & 0 \\
0 & 0 \\
0 & 0 \\
0 & 0 \\
0 & 0 \\
1 & 0 \\
0 & 1 \\
1 & 0
\end{bmatrix}
\quad
\begin{array}{c}
\Rightarrow \\
\text{Row 14} = \text{Row 1} + \text{Row 14} \\
\text{Row 15} = \text{Row 1} + \text{Row 15}
\end{array}
\quad
\begin{bmatrix}
-1 & -1 \\
1 & 0 \\
0 & 0 \\
0 & 0 \\
0 & 0 \\
0 & 1 \\
0 & 1 \\
0 & 0 \\
0 & 0 \\
0 & 0 \\
0 & 0 \\
0 & 0 \\
1 & 0 \\
-1 & 0 \\
0 & -1
\end{bmatrix}
$$

Adding row 1 of D_{uc} to eliminate the positive numbers in LD_{uc} corresponds to adding 1 to the first element of both rows of L to construct the new constraint matrix L'.

$$
L' = \begin{bmatrix}
0 & -1 & 0 & 0 & 0 & 0 & 0 & 0 & 0 & -1 & -1 & -1 & 0 \\
0 & 0 & 0 & 0 & -1 & -1 & 0 & 0 & 0 & 0 & -1 & -1 & 0
\end{bmatrix}
$$

$$
b' = \begin{bmatrix} -1 \\ -1 \end{bmatrix}
$$

The new constraint, $L'\mu_p \le b'$, represents the following inequalities:

$$\mu_2 + \mu_{10} + \mu_{c1} + \mu_{c2} \ge 1$$
$$\mu_5 + \mu_6 + \mu_{c1} + \mu_{c2} \ge 1$$

This inequality will insure that the number of tokens in S_1 remains positive and is also admissible with respect to the plant's uncontrollable transitions.

Note that with positive "less-than-or-equal-to" constraints, transforming an inequality usually results in adding new places to the constraint. Here the "greater-than-or-equal-to" nature of the inequalities resulted in the elimination of places.

The incidence matrix and initial marking of the control place are now calculated.

$$
D_c = -L'D_p = \begin{bmatrix}
-1 & 1 & 1 & 0 & -1 & 0 & 1 & 0 & 0 & 0 \\
-1 & 0 & 1 & 0 & 0 & 1 & 1 & 0 & -1 & 0
\end{bmatrix}
$$

$$
\mu_{c1_0} = b' - l'^T \mu_{p_0} = 1
$$

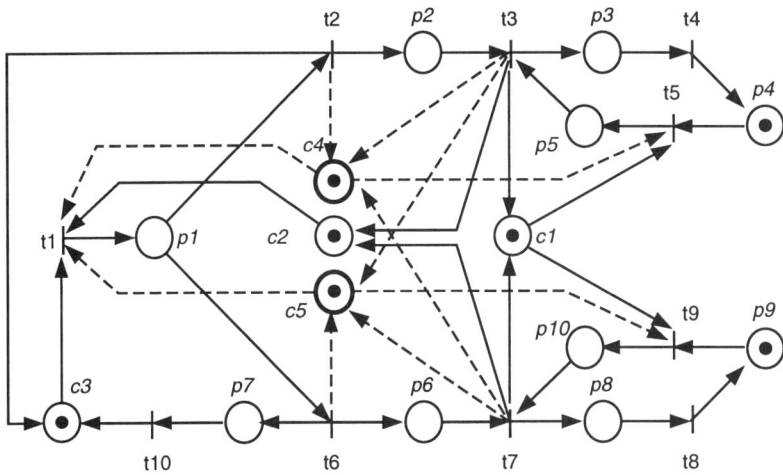

Figure 8.10.   The unreliable machine model is now live.

The controls for siphons S_1 and S_2 are shown as places c_4 and c_5 respectively in Figure 8.10. The supervised plant meets the constraints of section 8.3.2 and the controller will attempt no inhibition of uncontrollable transitions. Analysis of the closed loop system reveals that all of the net's siphons are controlled and the system is live.

8.4   Piston Rod Robotic Assembly Cell

8.4.1   Plant Definition and Constraints

The piston rod assembly cell application comes from chapter 8 of [Desrochers and Al-Jaar, 1995]. The Petri net model of the plant is shown in Figure 8.11. Table 8.2 details the meaning of each place in the net. A token in any of the Petri net places signifies that the action or condition specified in Table 8.2 is taking place. The piston rod assembly is performed by two robots, and the primary feedback mechanism is a vision system. An S-380 robot is used to prepare and align the parts for assembly, and an M-1 robot installs the cap on the piston rod. The specific duties of each robot are described below.

S-380: The S-380 robot remains idle until a new engine block and crank shaft become available. This event is represented by the appearance of a token in place p_1 in Figure 8.11. The firing of transition t_1 indicates the start of the process. At this time the S-380 moves the crank shaft into alignment and brings a new piston rod into the work area. These actions are represented by places p_2 and p_3. The firing of transition t_3 indicates that the S-380 has completed its duties for the particular engine block.

M-1: The M-1 robot starts its duties by picking up a piston pulling tool (place p_4) and, assuming the S-380 has brought a piston rod into position, pulls the piston rod into the engine block and replaces the pulling tool (place p_5). The M-1 then picks up a cap and secures it to the piston rod using two bolts (places p_6 and p_7). The firing of transition t_8 indicates that the M-1 has successfully installed the cap and the engine

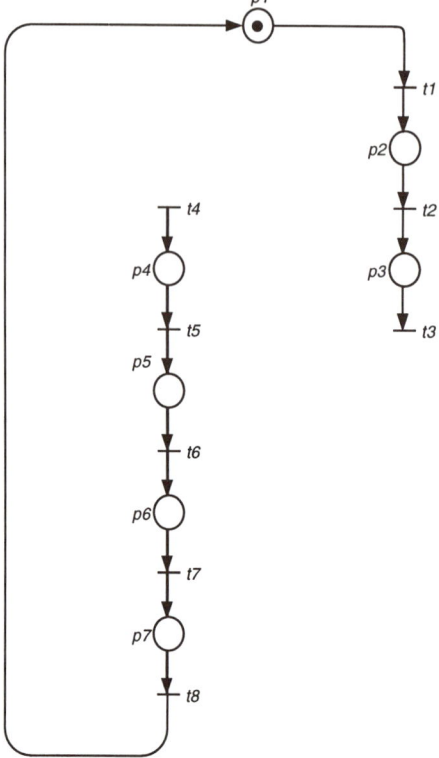

Figure 8.11.   The base Petri net model for the piston rod robotic assembly cell.

Table 8.2.   Place descriptions for Figure 8.11.

p_1	Work area clear, engine block and crank shaft ready to be processed.
p_2	S-380 robot aligns the crank shaft.
p_3	S-380 robot picks up new piston rod and positions it in work space.
p_4	M-1 robot picks up the piston pulling tool.
p_5	M-1 robot pulls the piston rod into the engine block, returns pulling tool.
p_6	M-1 robot picks up a cap and positions it on piston rod.
p_7	M-1 robot readies two bolts and uses them to fasten cap to piston rod.

block has been conveyed out of the work space. At this time work can begin on a new engine block.

The incidence matrix, D_p, and initial marking, μ_{p0}, of the plant are given by

$$D_p = \begin{bmatrix} -1 & 0 & 0 & 0 & 0 & 0 & 0 & 1 \\ 1 & -1 & 0 & 0 & 0 & 0 & 0 & 0 \\ 0 & 1 & -1 & 0 & 0 & 0 & 0 & 0 \\ 0 & 0 & 0 & 1 & -1 & 0 & 0 & 0 \\ 0 & 0 & 0 & 0 & 1 & -1 & 0 & 0 \\ 0 & 0 & 0 & 0 & 0 & 1 & -1 & 0 \\ 0 & 0 & 0 & 0 & 0 & 0 & 1 & -1 \end{bmatrix} \tag{8.10}$$

$$\mu_{p0} = \begin{bmatrix} 1 & 0 & 0 & 0 & 0 & 0 & 0 \end{bmatrix}^T$$

There are a number of constraints that must be imposed on this assembly cell in order to properly synchronize the robots and to insure that physical limitations are obeyed. There is only one S-380 and one M-1 robot available for the assembly process, thus only one of each robot can be working at any given time. Places p_2 and p_3 represent activity for the S-380 and places p_4, p_5, p_6, and p_7 represent activity for the M-1, thus

$$\mu_2 + \mu_3 \leq 1 \tag{8.11}$$

$$\mu_4 + \mu_5 + \mu_6 + \mu_7 \leq 1 \tag{8.12}$$

The plant graph already insures that the S-380 will not begin its task until an engine block and crank shaft are available and the work space is clear, however we need to make sure that the M-1 does not try to pull the piston rod into the engine until the S-380 has finished aligning the crank shaft and readying a new piston rod. In other words, we must insure that transition t_5 does not fire until after transition t_3 has fired. A review of the graph structure of the plant net shows that we can write this constraint as

$$\mu_1 + \mu_2 + \mu_3 + \mu_5 + \mu_6 + \mu_7 \leq 1 \tag{8.13}$$

Note that this constraint allows the M-1 robot to ready its pulling tool, the task of p_4, at the same time that the S-380 robot is performing one of its tasks.

There are several finite resources that are used in the assembly process. A piston rod is readied at p_3, a pulling tool is required at p_4 and p_5, a cap is required at p_6, and two nuts are used at p_7. The plant controller will need to know if any of these resources is unavailable in order to stall operation until the parts are ready. Section 6.1 shows that it is possible to provide the controller with the necessary hooks to manage these resources by associating a constraint with each of the places (or sets of places) that requires the use of a finite resource:

$$\mu_3 \leq 1 \tag{8.14}$$
$$\mu_4 + \mu_5 \leq 1 \tag{8.15}$$
$$\mu_6 \leq 1 \tag{8.16}$$
$$\mu_7 \leq 1 \tag{8.17}$$

These constraints are not intended to prohibit forbidden states, but rather to provide for the modeling of finite resource usage.

8.4.2   Uncontrollable Transitions and Control Synthesis

One final constraint placed on the assembly cell involves the smooth and uninterrupted operation of the M-1 robot. Perhaps due to the nature of the M-1 robot's dynamics or programming, we would prefer that its operation not be interrupted from the point that it pulls the piston rod into the engine block until it has completed fastening the cap on the piston rod. We can model this constraint by marking transitions t_6, t_7 and t_8 as uncontrollable. Thus we have told the controller that once a token passes into p_5, it can not stall the plant's progress until the token has passed on to places p_6, p_7 and completion. This means that if the controller does want to stall the M-1 robot, it should do it before it starts its primary operation.

Constraints $(8.11) - (8.17)$ are now combined into matrix form.

$$
\underbrace{\begin{bmatrix}
0 & 1 & 1 & 0 & 0 & 0 & 0 \\
0 & 0 & 0 & 1 & 1 & 1 & 1 \\
1 & 1 & 1 & 0 & 1 & 1 & 1 \\
0 & 0 & 1 & 0 & 0 & 0 & 0 \\
0 & 0 & 0 & 1 & 1 & 0 & 0 \\
0 & 0 & 0 & 0 & 0 & 1 & 0 \\
0 & 0 & 0 & 0 & 0 & 0 & 1
\end{bmatrix}}_{L}
\begin{bmatrix}
\mu_1 \\ \mu_2 \\ \mu_3 \\ \mu_4 \\ \mu_5 \\ \mu_6 \\ \mu_7
\end{bmatrix}
\leq
\underbrace{\begin{bmatrix}
1 \\ 1 \\ 1 \\ 1 \\ 1 \\ 1 \\ 1
\end{bmatrix}}_{b}
\qquad (8.18)
$$

The controller that enforces these constraints will have seven places (one for each row of L), but first we must insure that our constraints meet the uncontrollability requirements. Transitions t_6, t_7, and t_8 are uncontrollable, so the matrix D_{uc} is composed of the last three columns of D_p. According to chapter 5, the matrix LD_{uc} must be checked for any positive values, which would indicate that the current constraints may violate the uncontrollability conditions.

$$
LD_{uc} =
\begin{bmatrix}
0 & 0 & 0 \\
0 & 0 & -1 \\
0 & 0 & 0 \\
0 & 0 & 0 \\
-1 & 0 & 0 \\
1 & -1 & 0 \\
0 & 1 & -1
\end{bmatrix}
$$

Rows 6 and 7 of LD_{uc} contain positive values, thus it will be necessary to manipulate the sixth and seventh constraints (constraints (8.16) and (8.17)) so that their enforcement will not generate a controller with arcs directed toward the uncontrollable transitions. Based on the procedure outlined in chapter 5, the offending rows of LD_{uc} can be eliminated by adding rows from D_{uc}. Keeping track of the row operations performed will yield the matrices R_1 and R_2 which will be used to generate the transformed constraint matrix L'. The row operations are as follows.

$$\begin{array}{c}\text{Rows 6 and 7}\\\text{of } LD_{uc}\end{array} = \begin{bmatrix} 1 & -1 & 0 \\ 0 & 1 & -1 \end{bmatrix} \begin{array}{c}+\text{ row 5 of } D_{uc} \Rightarrow\\+\text{ row 6 of } D_{uc} \Rightarrow\end{array} \begin{bmatrix} 0 & -1 & 0 \\ 1 & 0 & -1 \end{bmatrix}$$

$$\Rightarrow\;\; +\text{ row 5 of } D_{uc} \Rightarrow \begin{bmatrix} 0 & -1 & 0 \\ 0 & 0 & -1 \end{bmatrix}$$

Now that the matrix no longer contains any positive numbers, we can use the row operations performed to create the matrices R_1 and R_2:

$$R_1 = \begin{bmatrix} 0 & 0 & 0 & 0 & 0 & 0 & 0 \\ 0 & 0 & 0 & 0 & 0 & 0 & 0 \\ 0 & 0 & 0 & 0 & 0 & 0 & 0 \\ 0 & 0 & 0 & 0 & 0 & 0 & 0 \\ 0 & 0 & 0 & 0 & 0 & 0 & 0 \\ 0 & 0 & 0 & 0 & 1 & 0 & 0 \\ 0 & 0 & 0 & 0 & 1 & 1 & 0 \end{bmatrix} \qquad R_2 = I$$

Constraints (8.16) and (8.17) have been transformed as follows:

$$\mu_6 \le 1 \;\Rightarrow\; \mu_5 + \mu_6 \le 1 \tag{8.19}$$
$$\mu_7 \le 1 \;\Rightarrow\; \mu_5 + \mu_6 + \mu_7 \le 1 \tag{8.20}$$

The new constraint matrix is given by

$$L' = R_1 + R_2 L = \begin{bmatrix} 0 & 1 & 1 & 0 & 0 & 0 & 0 \\ 0 & 0 & 0 & 1 & 1 & 1 & 1 \\ 1 & 1 & 1 & 0 & 1 & 1 & 1 \\ 0 & 0 & 1 & 0 & 0 & 0 & 0 \\ 0 & 0 & 0 & 1 & 1 & 0 & 0 \\ 0 & 0 & 0 & 0 & 1 & 1 & 0 \\ 0 & 0 & 0 & 0 & 1 & 1 & 1 \end{bmatrix}$$

and the incidence matrix and initial marking of the controller are calculated according to equations (3.10) and (3.11):

$$D_c = -L'D_p = \begin{bmatrix} -1 & 0 & 1 & 0 & 0 & 0 & 0 & 0 \\ 0 & 0 & 0 & -1 & 0 & 0 & 0 & 1 \\ 0 & 0 & 1 & 0 & -1 & 0 & 0 & 0 \\ 0 & -1 & 1 & 0 & 0 & 0 & 0 & 0 \\ 0 & 0 & 0 & -1 & 0 & 1 & 0 & 0 \\ 0 & 0 & 0 & 0 & -1 & 0 & 1 & 0 \\ 0 & 0 & 0 & 0 & -1 & 0 & 0 & 1 \end{bmatrix} \tag{8.21}$$

$$\mu_{c0} = \begin{bmatrix} 1 & 1 & 0 & 1 & 1 & 1 & 1 \end{bmatrix}^T \tag{8.22}$$

The controlled net is shown in Figure 8.12 with dashed controller arcs and thick-lined controller places. Table 8.3 describes the meaning of each of the controller places when a token is present within them. Note that place c_3 insures that the M-1 robot will wait to start working on the engine block until the S-380 has completed its task, but it is capable of readying the piston pulling tool while the S-380 is working. Also note that the controller directs no arcs to the uncontrollable transitions, however it still manages to enforce constraints (8.16) and (8.17).

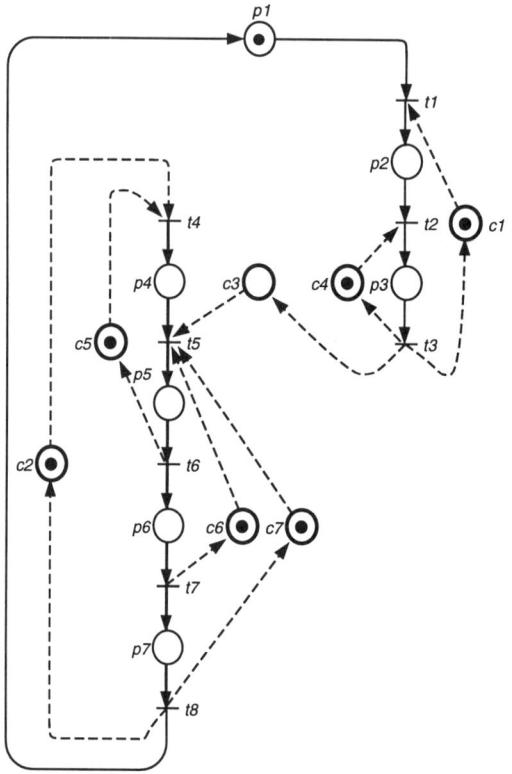

Figure 8.12.   The assembly cell model with Petri net controller.

Table 8.3.   Controller place descriptions for Figure 8.12.

c_1	S-380 robot is available for work.
c_2	M-1 robot is available for work.
c_3	S-380 robot has completed preparations, M-1 may begin.
c_4	A piston rod is available.
c_5	The piston pulling tool is available.
c_6	A cap is available.
c_7	Two nuts are available.

An examination of the enforced constraints represented by the matrix L' shows that some of these constraints are actually implied by others. For example, constraint (8.14) says that there can be at most one token in place p_3, but constraint (8.11) says that there can be at most one token, at any given time, in places p_2 and p_3. If constraint (8.11) is enforced then constraint (8.14) can not be violated, so why not remove constraint (8.14) as redundant? The answer is that constraint (8.14) and its associated controller place c_4 are not actually being used to prohibit a forbidden state, but to provide the controller with a means of dealing with finite resources, in this case, the availability of a new piston rod. This is the same situation with constraints (8.15), (8.16), and (8.17) and their associated places c_5, c_6, and c_7. The use of constraint inequalities for the modeling of finite resources is discussed in section 6.1.

8.4.3   Handling Sensor Failures

The piston rod assembly cell presented in [Desrochers and Al-Jaar, 1995] uses a vision system to provide sensory feedback to the controller. Suppose that an obstruction has appeared between the camera and the work space, partially obscuring the view of the M-1 robot's area. The controller can still observe the M-1 robot starting and completing its task, but it can no longer track the robot while it performs its duties. Transitions t_5, t_6, and t_7 have become unobservable (see section 4.2). This means that there should be no arcs from any of these transitions to the controller places, however it can be seen from (8.21) that the current version of the controller incidence matrix contains nonzero elements in columns five through seven. Let D_{uo} be a matrix composed of the unobservable columns of D_p. In order to compensate for the sensor failure, we will follow the procedure of section 5.2, starting by computing the kernel of D_{uo}. There are seven rows in D_{uo}, and the rank of the matrix is three. The kernel X will have $7 - 3 = 4$ rows.

$$
\underbrace{\begin{bmatrix} 1 & 0 & 0 & 0 & 0 & 0 & 0 \\ 0 & 1 & 0 & 0 & 0 & 0 & 0 \\ 0 & 0 & 1 & 0 & 0 & 0 & 0 \\ 0 & 0 & 0 & 1 & 1 & 1 & 1 \end{bmatrix}}_{X}
\underbrace{\begin{bmatrix} 0 & 0 & 0 \\ 0 & 0 & 0 \\ 0 & 0 & 0 \\ -1 & 0 & 0 \\ 1 & -1 & 0 \\ 0 & 1 & -1 \\ 0 & 0 & 1 \end{bmatrix}}_{D_{uo}} = 0
$$

Rows one through three of X tell us that constraints involving only μ_1, μ_2, and μ_3 are independent and will not have to be transformed in order to meet the unobservability requirements. However, row four of X indicates that the coefficients on μ_4, μ_5, μ_6, and μ_7 must be equal. It is now a simple matter to rewrite the set of constraints in order to meet this requirement:

$$\mu_2 + \mu_3 \leq 1 \Rightarrow \text{Unchanged}$$
$$\mu_4 + \mu_5 + \mu_6 + \mu_7 \leq 1 \Rightarrow \text{Unchanged}$$
$$\mu_1 + \mu_2 + \mu_3 + \mu_5 + \mu_6 + \mu_7 \leq 1 \Rightarrow$$
$$\mu_1 + \mu_2 + \mu_3 + \mu_4 + \mu_5 + \mu_6 + \mu_7 \leq 1$$
$$\mu_3 \leq 1 \Rightarrow \text{Unchanged} \tag{8.23}$$
$$\mu_4 + \mu_5 \leq 1 \Rightarrow \mu_4 + \mu_5 + \mu_6 + \mu_7 \leq 1$$
$$\mu_5 + \mu_6 \leq 1 \Rightarrow \mu_4 + \mu_5 + \mu_6 + \mu_7 \leq 1$$
$$\mu_5 + \mu_6 + \mu_7 \leq 1 \Rightarrow \mu_4 + \mu_5 + \mu_6 + \mu_7 \leq 1$$

In all cases, each transformed constraint can be realized by simple additions of new coefficients to the original constraints. This insures that the conditions of lemma 4.10 are obeyed: the new constraints will not allow states prohibited by the originals.

The constraints in (8.23) are now used to generate a reconfigured controller. The incidence matrix and initial marking are

$$D_c = \begin{bmatrix} -1 & 0 & 1 & 0 & 0 & 0 & 0 & 0 \\ 0 & 0 & 0 & -1 & 0 & 0 & 0 & 1 \\ 0 & 0 & 1 & -1 & 0 & 0 & 0 & 0 \\ 0 & -1 & 1 & 0 & 0 & 0 & 0 & 0 \\ 0 & 0 & 0 & -1 & 0 & 0 & 0 & 1 \\ 0 & 0 & 0 & -1 & 0 & 0 & 0 & 1 \\ 0 & 0 & 0 & -1 & 0 & 0 & 0 & 1 \end{bmatrix} \tag{8.24}$$

$$\mu_{c0} = \begin{bmatrix} 1 & 1 & 0 & 1 & 1 & 1 & 1 \end{bmatrix}^T$$

Observe how the fifth through seventh columns of D_c have been completely zeroed. The reconfigured control is shown in Figure 8.13.

The last three transformed constraints in (8.23) are identical, however all constraints will be implemented separately in the reconfigured version of the controller. This is because the different controller places have different interpretations (see table 8.2) and because we intend to transform the controller arcs back to their original configurations once the sensor obstruction has been removed.

8.5   Asynchronous Transfer Mode Switch

This section provides an illustration of the graph transformation technique described in section 7.2.1 used for handling constraints on allowable events (as opposed to allowable states). The example involves the control logic of the output-port controllers (OPC) of an asynchronous transfer mode (ATM) switch. The plant description is taken from [Li and Wonham, 1995]. An ATM switch is used to route segments of a 53 byte data stream referred to as ATM cells. Incoming cells are directed by a series of routing networks to the OPC's. Each OPC holds the outgoing cells in a buffer as they wait to be transmitted. If an OPC buffer is full then any further cells that the routing network might send to the OPC are lost until a new space is cleared in the buffer. The OPC is

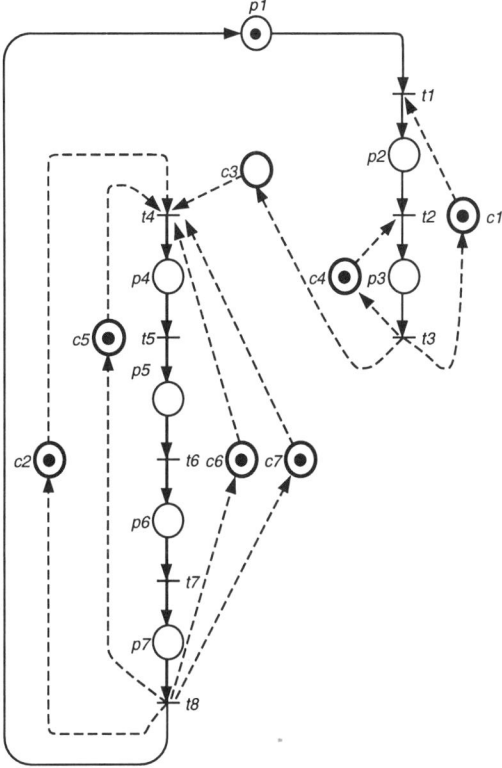

Figure 8.13.   The assembly cell model with a controller that accounts for a sensor loss making transitions t_5, t_6 and t_7 unobservable.

capable of inhibiting the transmission of further cells from the routing networks if its buffer is too full. The transmission of data in the ATM switch is extremely fast and the ATM cells are often independent of each other; for these reasons the output-port controllers' buffers are capable of receiving more than one cell at a time, however we will assume here that the buffers may lose data if an attempt is made to simultaneously deliver more than two cells to any OPC's buffer.

A Petri net model of the portion of interest of the ATM switch is shown in Figure 8.14. Three routing networks are represented by p_1, p_2, and p_3. These networks receive ATM cells through the uncontrollable firings of t_1, t_2, and t_3. Places p_4 and p_5 are the output buffers. Transitions t_{10} and t_{11} fire to represent the transmission of ATM cells in the buffers. From the point of view of the OPC, these transitions are uncontrollable, i.e., if its buffer is running out of room, it can not clear space by forcing the transmission of a cell in the buffer, instead it must inhibit the transmission of further cells from the routing networks. Each OPC buffer has a capacity of five cells, thus the first two constraints are

$$\mu_4 \leq 5 \qquad\qquad (8.25)$$

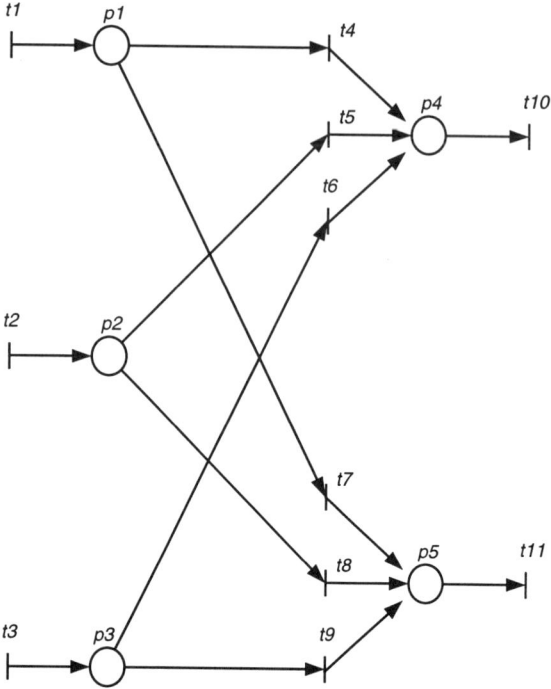

Figure 8.14.   Petri net model for the output portion of an ATM switch.

$$\mu_5 \leq 5 \tag{8.26}$$

Data may be lost if more than two ATM cells enter the OPC buffer at any one time, so the final two constraints are constraints on the firing vector:

$$q_4 + q_5 + q_6 \leq 2 \tag{8.27}$$
$$q_7 + q_8 + q_9 \leq 2 \tag{8.28}$$

To use the controller generation procedure of chapter 3 we first perform the transformation of all transitions involved in the constraints as described in section 7.2.1. The result is shown in Figure 8.15. The transformation permits constraints (8.27) and (8.28) to be written

$$\mu'_4 + \mu'_5 + \mu'_6 \leq 2 \tag{8.29}$$
$$\mu'_7 + \mu'_8 + \mu'_9 \leq 2 \tag{8.30}$$

The controller is allowed to take direct action to prevent any OPC buffer from receiving too many ATM cells at a time, so the procedure for the direct enforcement of firing vector constraints from section 7.2.1 is used to enforce (8.29) and (8.30). After synthesizing the controller for the four constraints, ((8.25),(8.26),(8.29),(8.30)), the dummy places $(p'_4, \ldots, p'_9)$ and the dummy transitions $(t'_4, \ldots, t'_9)$ of Figure 8.15 are

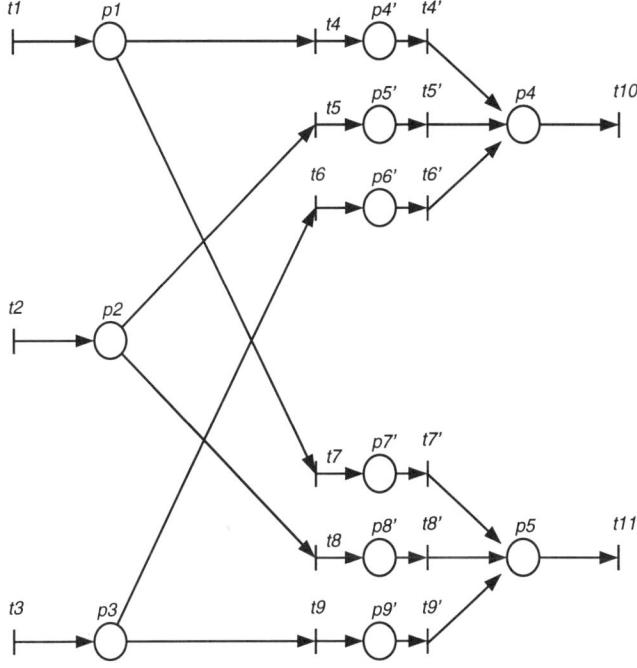

Figure 8.15.   The Petri net of Figure 8.14 with dummy places and transitions to account for firing vector constraints.

collapsed to put the plant model back into its original form. The resulting controlled system is shown in Figure 8.16. Control places c_1 and c_2 are used to prohibit further transmission from the router networks when an OPC buffer is full (five ATM cells). Places c_3 and c_4 only allow two transition firings at any given time to a single output buffer using the inherent rules for the handling of concurrent firings for Petri net models.

8.6   The Three Tanks Problem

The plant of Figure 8.17 consists of three fluid-filled tanks and a pump that can add fluid to each of them [Labinaz et al., 1997, Chase et al., 1993]. Tank i drains at a constant rate d_i. The pump can be moved to fill any of the three tanks. It fills tanks at a constant rate F. The control goal is to make sure that the fluid level in each of the three tanks stays within a safe boundary area. The fluid in tank i is not to rise above h_i nor is it to drain below l_i. It is possible, by switching the pumps to different tanks, to meet these constraints by making the following three assumptions:

1. The time required to move the pump from a tank to any other tank is 0.

2. The net flow out of the tanks is equal to the net flow in:

$$F = d_1 + d_2 + d_3 \qquad (8.31)$$

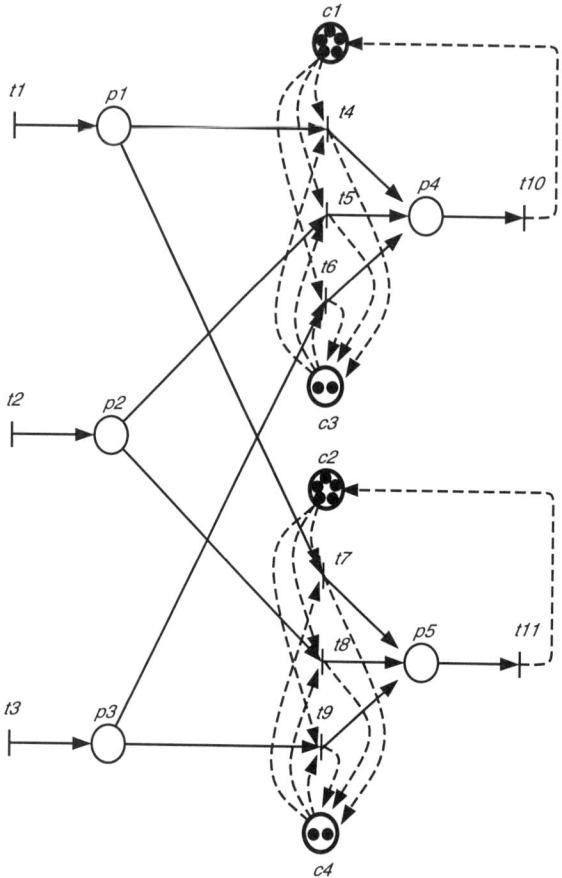

Figure 8.16.   The ATM switch with output-port control logic in place.

3. The initial fluid levels are well within the safety ranges.

These assumptions are not too unrealistic. Assumption 2 can be established by using a feedback controller to regulate F such that it is equal to the net flow out of the tanks. Assumption 1 can be eliminated by modifying assumption 2 such that F accounts for the time required to move from tank to tank as well as the net outflow. For the purposes of this discussion, and for supervisory control, we will leave the assumptions as they are written.

There are three actions that can be made by the controller, u_1, u_2, u_3, where $u_i \in \{0, 1\}$ and

$$u_i = \begin{cases} 1 & \text{Move to tank } i \text{ and fill.} \\ 0 & \text{No command regarding tank } i. \end{cases}$$

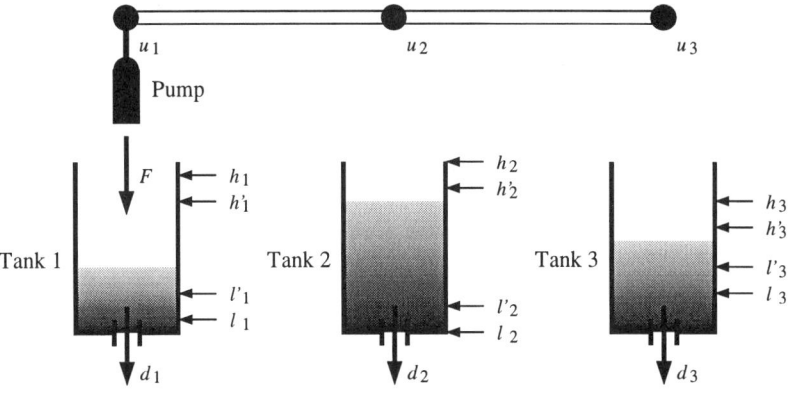

Figure 8.17.   Three fluid filled tanks serviced by a single mobile pump.

Because the pump can only service one tank at a time, we have

$$u_1 + u_2 + u_3 \leq 1$$

The pump is always filling one of the tanks, so the control-action constraint could actually be written $u_1 + u_2 + u_3 = 1$ if we assume that the command "Move pump to tank i" has no effect on the system when the pump is already servicing tank i. This assumption will be made in the discussion that follows.

The controller's feedback measurements of the tanks' situation comes from periodic samplings. The sampling period, T, and the fill rate, F, are then combined in order to find pseudo upper (h') and lower (l') bounds for the fluid levels in the tank.

$$h_i' \; \leq \; h_i - FT \tag{8.32}$$
$$l_i' \; > \; l_i + 2FT \tag{8.33}$$

These bounds mean that if tank i is currently being filled by the pump, and its fluid level rises above h_i', then the tank may overflow past h_i during the next sampling period unless the pump moves immediately to a different tank. The lower bounds, l_i', are constructed similarly, except that two sampling periods are allowed before the pump must respond in order to allow for the possibility that two tanks may reach a dangerously low level during the same sampling period. Finally, a bound is placed on the minimum distance between h_i' and l_i', insuring that each tank has an achievable safety region:

$$h_i' > l_i' + FT$$

With the given assumptions and control actions, the constraint to keep the fluid levels within the h' and l' bounds can be met by implementing the following control law:

1. Move to tank i when its fluid level drops below l_i'

2. Move to tank $j \neq i$ when the fluid level in tank i rises above h_i'.

Nothing is specifically stated about the dynamics of the flowing fluid. The inputs coming into the controller are events indicating fluid levels, and the control output is one of three discrete commands. For these reasons, the problem can be handled with DES supervisory control. Two approaches to designing the controller are investigated in the following sections.

8.6.1   Forbidden State Elimination

Supervisory controllers that use automata are generally derived by creating an automaton model of the plant, and deleting states from the model such that the modified model speaks a desired language. The controller automaton is then equivalent to this modified model. Controllable and uncontrollable events mark the passage from state to state in the automaton. If a set of forbidden states can be identified and then extracted along with all the states that might lead uncontrollably to the forbidden states, then the controller is equal to this reduced plant model. At any stage in the evolution of the plant, the controller automaton is consulted in order to determine the set of valid controllable actions that may be taken at that time.

In order to model one tank in the fluid-filled tank problem, we can use a five state automaton. Let x be the fluid level in the tank, then the five states are

1. Overflow: $x \geq h$

2. Pseudo overflow $h' \leq x < h$.

3. Fluid level OK: $l' < x < h'$.

4. Pseudo underflow $l < x \leq l'$.

5. Underflow: $x \leq l$

The tank starts in state 3, according to the assumptions. The action of pumping fluid into the tank will, given enough time, result in the transition from higher numbered states to lower numbered states. When no fluid is pumped into the tank, transitions from lower numbered states to higher numbered states will take place.

A complete automaton model of the three tank system would have $5^3 = 125$ states. A controller could be constructed by creating a duplicate automaton and then removing the states that involve over/underflow of any of the tanks. At any point, the state of the controller could then be used to determine which tanks may (or must) have fluid pumped to them at any given time.

A similar approach can be taken using Petri nets. A Petri net model of the three tank system is shown in Figure 8.18. The model consists of three unconnected nets representing each tank. For tank 1, place i is equivalent to state i in the single tank automaton described above. Since place three is marked, this means that, currently, the fluid level in tank 1 is within acceptable bounds: $l'_1 < x_1 < h'_1$. Note that the Petri net model of the problem is superior to the automaton in terms of space used by the model. The Petri net model grows polynomially with the number of tanks, while the automaton grows exponentially.

The transitions in Figure 8.18 would be fired in unison with observed events in the real system. The transitions are labeled with the control actions that could have been

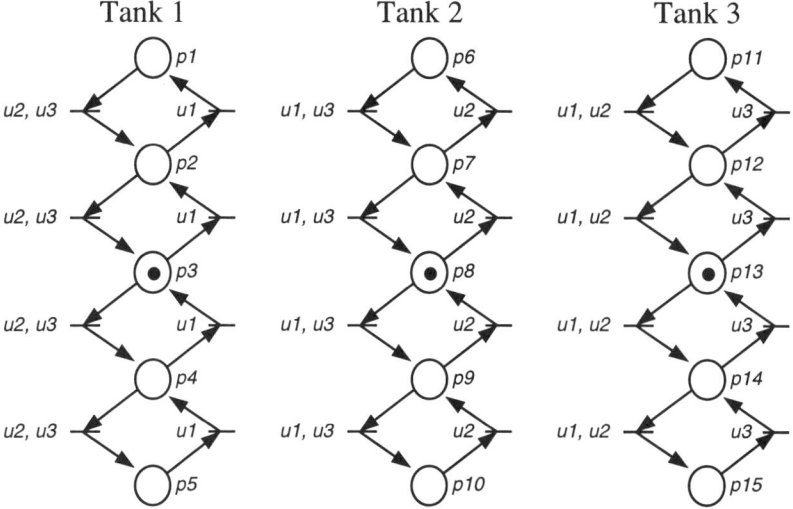

Figure 8.18.   A Petri net model of the discrete state space for the fluid-filled tank problem.

in effect at the time that the transition occurred. For example, the token in p_3 can only move to p_2 if the pump is servicing tank 1, so the transition is labeled u_1. Similarly, the token in p_3 can only move to p_4 when either tank 2 or tank 3 is being serviced, so this transition is labeled u_2, u_3.

In order to create a controller, the forbidden states are prevented from containing any tokens:

$$\mu_1 = 0$$
$$\mu_5 = 0$$
$$\mu_6 = 0$$
$$\mu_{10} = 0$$
$$\mu_{11} = 0$$
$$\mu_{15} = 0$$

The Petri net model of the controller is then given in Figure 8.19. A constraint of the form $\mu_i = 0$ can be implemented using the standard PN control techniques on the equivalent constraint $\mu_i \leq 0$, but the net in Figure 8.19 simply shows empty places added to the net to disable the appropriate transitions.

The controller net is used as follows. Observations of the plant will lead to transition firings in the controller net. After any transition firing, determine which transitions are disabled at every place in the net that contains a token. The labels on these transitions ($u_1, u_2,$ or u_3) indicate control actions that can not be taken at this time. Choose a control action from those that remain.

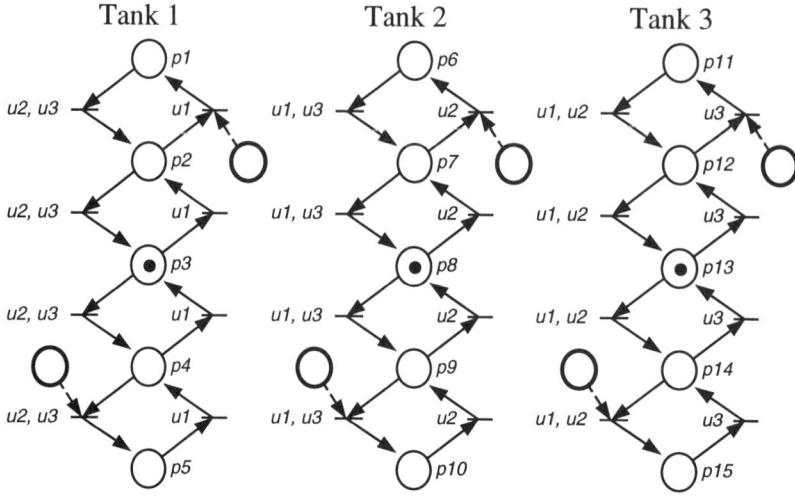

Figure 8.19.   The fluid-filled tank model with entrances to forbidden states disabled.

The approach above has several problems. It does not really take advantage of the analysis and control-synthesis powers of Petri nets. It is little more than a lookup table modeled as a Petri net. The Petri net implementation of the controller does involve a smaller model than a single automaton model of the system, but there is nothing preventing a control designer from using three separate automata and then combining their control action indicators as was done with the Petri nets above. In this situation, the models, PN or automata, would require the same amount of space and would be virtually identical.

Another problem with the solution presented here is that it may lead to deadlock. Suppose two tanks drained below their pseudo lower bound during the same sampling time. The choice of available control actions would then be the empty set if the description of how to choose the control actions above were taken literally. This problem could be overcome by using careful rules which determine how the states in the net are allowed to evolve, but instead a more graceful method of dealing with the problem is presented in the next section. There the control is designed in a way that takes advantage of the dynamic properties of Petri nets and the ability to automatically enforce constraints on them.

8.6.2   Event Driven

There are three states of interest with regard to any tank:

1. Tank fluid level is greater than pseudo overflow.

2. Tank fluid level is OK.

3. Tank fluid level is less than pseudo underflow.

There are four relevant events associated with the evolution of any given tank:

1. Tank fluid level has raised past pseudo overflow.

2. Tank fluid level has lowered below pseudo overflow.

3. Tank fluid level has lowered below pseudo underflow.

4. Tank fluid level has raised above pseudo underflow.

These events are uncontrollable, they occur upon observation of processes in the plant. There are three controllable events, $u_1, u_2,$ and u_3, associated with the placement of the pump. These states and events are combined to form the Petri net model of Figure 8.20. The meaning of each place and transition within the model is defined in Table 8.4.

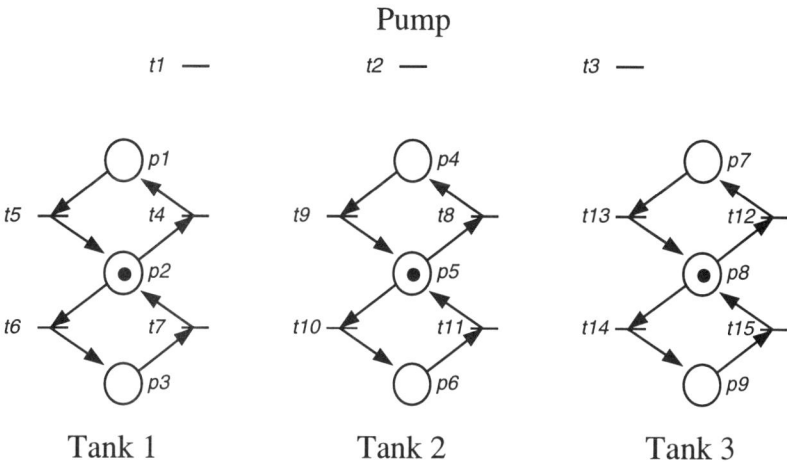

Figure 8.20.   A Petri net model of the relevant events and states for the fluid-filled tank problem.

Using the base model of the system, we will synthesize a controller Petri net such that the control $(u_1, u_2,$ or $u_3)$ issued to the plant is equivalent to the firings of $t_1, t_2,$ or t_3. The first step is to determine the constraints on the plant model. The first constraint is induced by the physical realities of the plant: only one pump can be serviced at a time.

$$q_1 + q_2 + q_3 \leq 1 \tag{8.34}$$

Transitions 1, 2, and 3 are all controllable, so the direct enforcement (section 7.2.1) of constraint (8.34) is admissible by Corollary 7.2 and may be implemented.

We need to insure that the tanks will not overflow, this means that we can not deliver fluid to tank i when tank i is experiencing pseudo overflow.

$$q_1 + \mu_1 \leq 1 \tag{8.35}$$
$$q_2 + \mu_4 \leq 1 \tag{8.36}$$
$$q_3 + \mu_7 \leq 1 \tag{8.37}$$

Table 8.4.   Explanation of transitions and places in Figure 8.20.

Controllable transitions	
t_1	Move pump to tank 1 and fill.
t_2	Move pump to tank 2 and fill.
t_3	Move pump to tank 3 and fill.

Uncontrollable transitions	
t_4, t_8, t_{12}	Pseudo overflow occurred.
t_5, t_9, t_{13}	Pseudo overflow cleared.
t_6, t_{10}, t_{14}	Pseudo underflow occurred.
t_7, t_{11}, t_{15}	Pseudo underflow cleared.

Places	
p_1, p_4, p_7	Pseudo overflow.
p_2, p_5, p_8	Fluid level OK.
p_3, p_6, p_9	Pseudo underflow.

Before implementing these constraints, we must first determine that they are admissible. This procedure starts by determining if each constraint is *uncoupled* (see Definition 7.3 and Proposition 7.4).

For constraint (8.35), the set T_f corresponds to the q_1 portion of the constraint and T_l corresponds to μ_1 portion of the constraint. Using the rules for controller construction and/or Proposition 7.4 we see that $T_f = \{t_1\}$ and $T_l = \{t_4, t_5\}$, thus $T_l \cap T_f = \emptyset$ and the constraint is uncoupled. The same can be shown for constraints (8.36) and (8.37). Given that they are uncoupled, the admissibility of the three constraints can be determined with Proposition 7.5.

Proposition 7.5 indicates that if the two constraints $\mu_1 \leq 1$ and $q_1 \leq 1$ are admissible, then (8.35) is admissible. Corollary 7.2 shows that $q_1 \leq 1$ is admissible, since the transition it effects (t_1) is a controllable transition. Next the admissibility of $\mu_1 \leq 1$ must be verified.

A structural analysis of the plant indicates the presence of the following place invariant:

$$\mu_1 + \mu_2 + \mu_3 = 1 \tag{8.38}$$

Since equation (8.38) is always true throughout the evolution of the plant, $\mu_1 \leq 1$ is always true, which implies that $\mu_1 \leq 1$ is admissible according to Corollary 4.8. Having determined that $q_1 \leq 1$ and $\mu_1 \leq 1$ are both admissible, we have satisfied both conditions in Proposition 7.5 and have demonstrated that constraint (8.35) is indeed admissible. *The controller associated with this constraint will direct an arc to the uncontrollable transition t_4. However Proposition 7.5 and Corollary 4.8 insure that this arc will only be used for observation and never for inhibition.* A similar application of Proposition 7.5 indicates that constraints (8.36) and (8.37) are also admissible.

Finally we need to prevent underflow, which means that the pump must shift to the tank that has sent a "pseudo-underflow occurred" event. Supervisory control is not equipped to make a direct demand that a particular transition be fired. Instead, during conditions of possible underflow, we will disable all actions that do not help to solve the problem. The designer is free to choose rules under which enabled transitions fire. If we state that one of the enabled transitions, t_1, t_2, or t_3, must fire every sampling period, then disabling undesirable actions is equivalent to forcing desirable actions.

Consider the situation where we wish to disable flow to the first tank. Clearly if tank 1 is in a safe region, and the fluid level in tank 2 has lowered below l_2', then we wish to disable t_1 so that the pump will no longer have the option of filling tank 1 and be able to move to tank 2. This constraint is expressed

$$q_1 + \mu_6 \leq 1$$

Thus t_1 may not fire when tank 2 requires service (indicated by the presence of a token in p_6). We also want to disable t_1 if tank 3 is experiencing underflow. The constraint is now modified:

$$2q_1 + \mu_6 + \mu_9 \leq 2 \tag{8.39}$$

If both tank 2 and 3 are above their underflow levels, then t_1 will be free to fire, but as soon as either of them drops below this level, t_1 will be disabled. But now suppose tank 1 is in danger of underflowing. Under no circumstances do we wish to disable t_1 if p_3 contains a token. Thus the constraint is modified a final time:

$$2q_1 + \mu_6 + \mu_9 - 2\mu_3 \leq 2 \tag{8.40}$$

Now if p_3 is clear, the constraint will behave exactly as in (8.39), but if p_3 contains a token, then the modified constraint will always be true, independent of the values of μ_6 and μ_9, and q_1 will be free to fire. The remaining underflow-prevention constraints are

$$2q_2 + \mu_3 + \mu_9 - 2\mu_6 \quad \leq \quad 2 \tag{8.41}$$
$$2q_3 + \mu_3 + \mu_6 - 2\mu_9 \quad \leq \quad 2 \tag{8.42}$$

The admissibility of these constraints can be verified using proposition 7.5.

Constraints (8.39)–(8.42) allow for tanks to simultaneously fall below their pseudo lower bounds without causing deadlock, since t_i will always be enabled when tank i is in the underflow state. Constraint (8.34) will insure that the controller only tries to service one of these at any given time. Assumption (8.31) insures that at most only two tanks can experience pseudo underflow at any given time. Assumptions (8.31) and (8.33) show that the first tank to receive service will go above its l' level in one sampling period, leaving time to fill the remaining tank before it drops below its l level.

A controller for enforcing the constraints, (8.34),(8.35)–(8.37),(8.39)–(8.42), is automatically generated using the rules for direct implementation of firing vector constraints (section 7.2.1). The controlled plant is shown in Figure 8.21. Note that even though the diagram is fairly complex, the control places and arcs were all automatically generated. The plant is controlled through simple simulation of the Petri net's evolution. Actual observed events from the plant are used to determine when the plant transitions fire. The control actions (t_1, t_2, and t_3) fire every sampling period. Any method can be used to determine which of the three should fire when there is a choice.

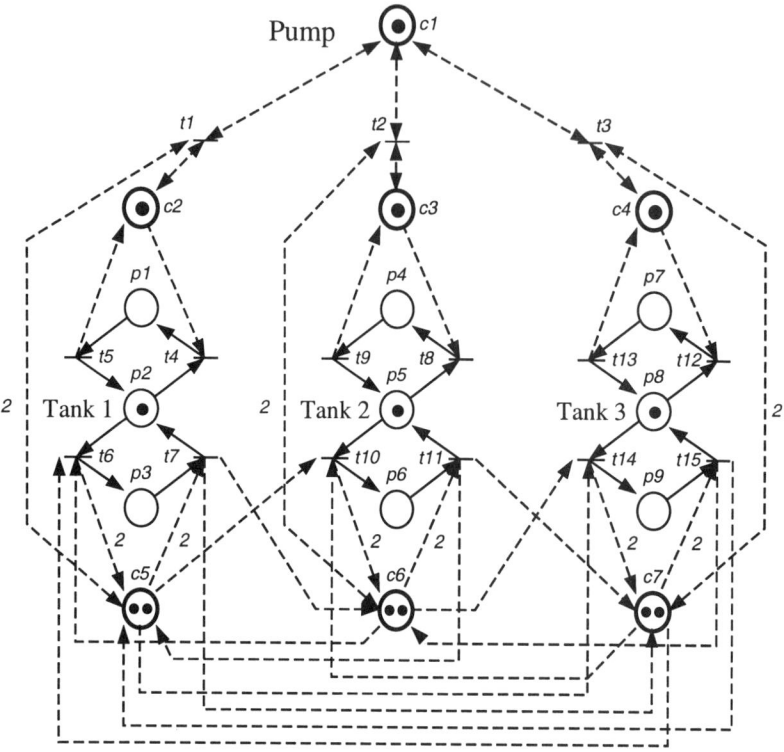

Figure 8.21.   The plant of Figure 8.20 with added control structures.

8.7 Hybrid Control System Example

The plant of this example is a simple linear system that has been used extensively by [Stiver, 1995, Antsaklis et al., 1993b, Stiver et al., 1996, Lemmon et al., 1993] to demonstrate concepts of hybrid control. The continuous plant will be modeled as a discrete event system and its behavior will be regulated using the control techniques described in chapter 3 and section 7.2.1.

The continuous plant model is described by

$$\dot{x}(t) = f(x, u) = \begin{bmatrix} 0 & 1 \\ 0 & 0 \end{bmatrix} x(t) + \begin{bmatrix} 0 \\ 1 \end{bmatrix} u(t) \tag{8.43}$$

where t is time, $x(t) = [x_1(t)\ x_2(t)]^T$ is the state vector, and $u(t)$ is the control input. The plant is referred to as a double integrator since x_2 is the integral of the input and

x_1 is the integral of x_2. The control goal is to drive the state variables from their initial conditions to a unit disk centered at the origin.

The dynamics of the plant are represented as a DES by dividing the state space into several regions and assigning Petri net places to each of these regions. The borders between the regions are defined by a set of hypersurfaces. The first hypersurface, h_1, represents the target region:

$$h_1(x) = x_1^2 + x_2^2 - 1 \tag{8.44}$$

thus the state is within the target region when $h_1(x) < 0$. In order to define the other surfaces we will restrict the allowed inputs so as to limit the possible number of region-to-region transitions. Let $u(t)$ be a piecewise constant input with values ± 1. A desirable partitioning will yield a state transition function that indicates a unique next state given the current state and control input. Failure to meet this requirement will yield a nondeterministic plant DES, which is difficult to use as a model for control. Following the procedure detailed in [Stiver, 1995], an acceptable partitioning of the continuous state space will be found by placing the remaining hypersurfaces along invariant plant trajectories, i.e., the hypersurfaces will satisfy the equation

$$f(x) \cdot \nabla h(x) = 0$$

Choosing the input $u(t) = 1$ we have $f(x) = [x_2 \ 1]^T$ and

$$f(x) \cdot \nabla h(x) = x_2(\nabla h)_1 + (\nabla h)_2 = 0$$

To satisfy the equation we choose

$$\nabla h = \begin{bmatrix} -1 \\ x_2 \end{bmatrix} \tag{8.45}$$

Equation (8.45) is integrated to find a solution for $h(x)$:

$$\begin{aligned} h(x) &= \int -1 dx_1 = \int x_2 dx_2 \\ &= -x_1 + g_1(x_2) = \frac{1}{2}x_2^2 + g_2(x_1) \\ \Rightarrow h(x) &= \frac{1}{2}x_2^2 - x_1 + c \end{aligned} \tag{8.46}$$

where c is a constant. Performing a similar procedure for $u(t) = -1$ yields

$$h(x) = \frac{1}{2}x_2^2 + x_1 + c \tag{8.47}$$

Two new partitioning hypersurfaces are now formed by creating tangents to the convex region described by (8.44) from the invariant surfaces (8.46) and (8.47):

$$h_2 = \begin{cases} \frac{1}{2}x_2^2 + x_1 + 1 & ; \quad x_2 \geq 0 \\ \frac{1}{2}x_2^2 - x_1 - 1 & ; \quad x_2 < 0 \end{cases} \tag{8.48}$$

$$h_3 = \begin{cases} \frac{1}{2}x_2^2 + x_1 - 1 & ; \quad x_2 \geq 0 \\ \frac{1}{2}x_2^2 - x_1 + 1 & ; \quad x_2 < 0 \end{cases} \tag{8.49}$$

$$\tag{8.50}$$

The surfaces where chosen so as to form regions of the state space that lead invariably toward the target region under the proper constant input. Figure 8.22 shows the three hypersurfaces as dotted lines with the corresponding regions numbered 1 – 5. The properties of each region are as follows.

Region 1 Approach target region with $u = -1$.

Region 2 Approach target region with $u = +1$.

Region 3 Approach region 1 with $u = +1$.

Region 4 Approach region 2 with $u = -1$.

Region 5 Target.

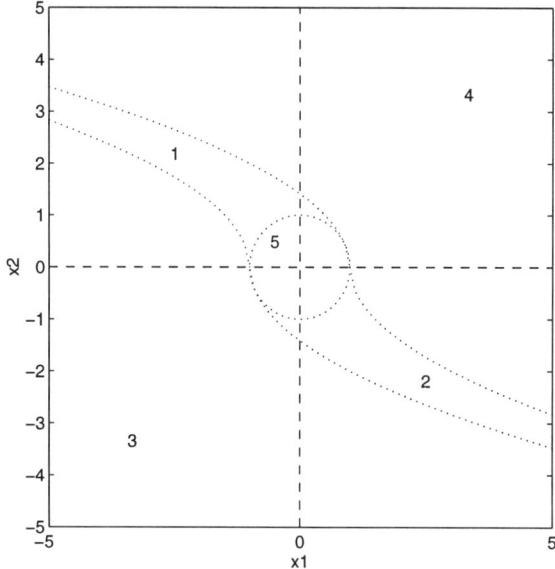

Figure 8.22.   Two dimensional continuous state space with three hypersurfaces and five discrete regions.

A Petri net model of the plant is constructed and shown in Figure 8.23. Place p_n in the figure corresponds to region n of the state space. The number in parentheses next to each transition corresponds to the value of u that can cause the transition to occur. Transitions t_5 and t_6 do not actually represent state transitions; they are included because they represent undesirable input values that can cause the continuous trajectory of the plant to wander far from the target region near the origin. Because the state space regions do not overlap and the transitions are deterministic with respect to the input, this Petri net is a state machine. The Petri net structure is useful because it will allow us to automatically generate a supervisor based on a set of linear constraints.

The first constraint deals with the undesirable behavior described by t_5 and t_6. We will disable these transitions:

$$q_5 + q_6 \leq 0 \qquad (8.51)$$

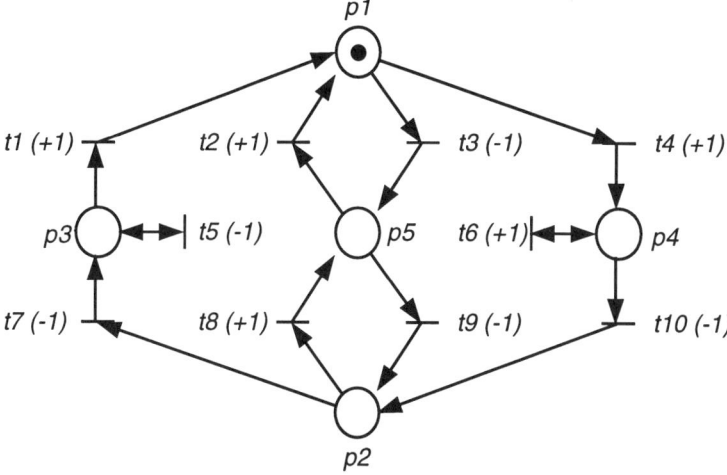

Figure 8.23.   Petri net plant model for the double integrator.

The second constraint indicates that we wish the trajectory to remain within the region of the target area:

$$\mu_1 + \mu_2 + \mu_5 \geq 1 \tag{8.52}$$

This constraint poses a slight difficulty in that it is possible the condition imposed by the constraint will not be met by the initial conditions of the plant, e.g., if the initial condition lies within region 3 of the state space, constraint (8.52) will yield a control place acting as an excess variable with an initial marking of -1. Negative markings are invalid for Petri nets, however, in the spirit of more general "vector discrete event systems" [Li and Wonham, 1993] and the modified control structures of section 5.4, if this initial negative marking is allowed and proper Petri net rules are obeyed after the initialization, the controller will still work. Another approach to the implementation of constraint (8.52) is to wait and enforce it dynamically after the plant trajectory has progressed into a state where (8.52) is true. In this situation, the plant would progress naturally, obeying constraint (8.51), until the token entered place p_1 or p_2, at which time the token would be trapped by the addition of a controller for (8.52).

The Petri net supervisor for this problem is constructed using the method given in chapter 3 after performing the transformation of constraint (8.51) using the procedure for the direct enforcement of firing vector constraints described in section 7.2.1. The resulting controlled net is shown in Figure 8.24. Figure 8.25 shows two example trajectories of the controlled plant through the continuous state space. The initial conditions of each trajectory are marked with $X's$ on the graph. The figure shows the trajectories proceeding randomly after they enter the target region in order to emphasize that the Petri net controller is a supervisor: it does not inhibit or direct the behavior of the system as long as the state remains sufficiently close to the origin.

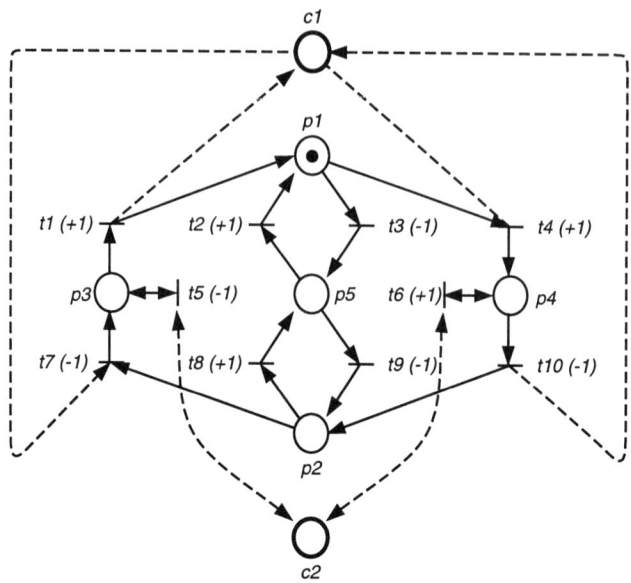

Figure 8.24.   Double integrator plant model with supervisor.

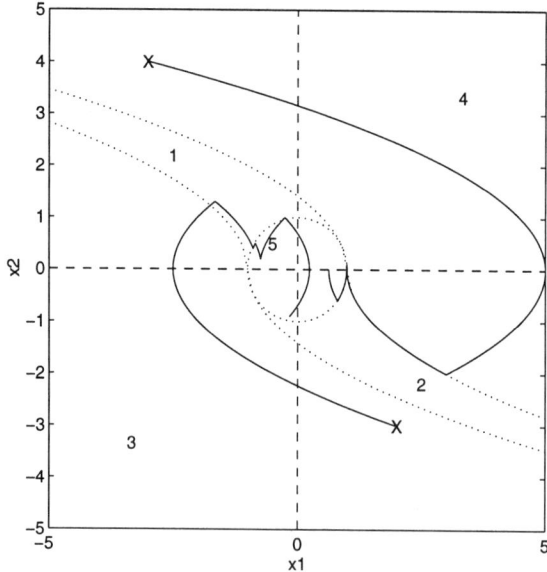

Figure 8.25.   Example trajectories of the supervised double integrator.

9 SUMMARY AND CONCLUSIONS

Petri nets possess many assets as models for discrete event systems. Concurrent processes and events can be easily modeled within the framework. They provide for larger reachable state spaces, more compact representation, and increased behavioral complexity compared to automata based models. The goal of this book has been to present an approach to Petri net supervisory control that is unified and tractable as well as comprehensive and practical.

The primary technical tools required for the use and analysis of the control methods presented here involve Petri net theory and matrix algebra. The techniques for deadlock avoidance and liveness insurance are derived from net theory. The main synthesis technique is based on the idea that specifications representing desired plant behavior can be enforced by making them invariants of the closed loop system. Most of the other tools in this book also revolve around the creation or characterization of invariants or an analysis of the interrelation between control specifications and plant and controller structure. For example, the methods for handling uncontrollable and unobservable transitions are motivated by observing the nature of the arcs induced in a controller by a given plant and specification.

Because an invariant based controller is itself a Petri net, the unified plant / controller system facilitates the use of established synthesis and analysis methods. The closed loop system can be designed, analyzed, simulated, verified, and augmented using tools already established for Petri nets. The particular approach to uncontrollable and unobservable transitions in this book was made possible by an initial understanding of the closed loop system as a Petri net.

The control method is able to cope with the most difficult and important aspects of supervisory control: uncontrollable events, unobservable events, and deadlock. Using the idea that the supervisor is modeled by an external Petri net has led to a novel approach to handling uncontrollability. Unobservable transitions have not received equal attention in the DES control literature, but they present an important problem, and systems that incorporate unobservable events are of practical concern. Here the problem of unobservability has been presented and analyzed concurrently and analogously to uncontrollability.

Methods have been presented for characterizing all feasible invariant based controllers for enforcing a linear place-constraint on a plant with uncontrollable and unobservable transitions. This characterization can be combined with an extended PN controller definition to enforce the logical union of all these feasible controls. Supervisors designed this way will allow a high degree of plant freedom. Unfortunately it has not been demonstrated that these controllers will always be maximally permissive, since it is not known if there is a situation where the maximally permissive control law corresponding to a linear predicate is ever something other than a disjunction of other linear predicates.

Supervisors may allow or even induce local or complete deadlock in a plant if the problem of liveness is not specifically dealt with in the supervisor design. The incorporation of established deadlock avoidance procedures into the invariant based control technique greatly expands the applicability and practical use of this method. At the same time, the techniques for deadlock avoidance benefit from the added ability to handle uncontrollable and unobservable transitions as well as a variety of other system constraints.

Aside from the standard forbidden state avoidance problems, invariant based controllers have been shown to be useful for a number of other supervisory specifications. The allocation and use of finite resources can be modeled and controlled as if a forbidden state problem were being solved. The addition of PN clock structures to the net allows for real-time specifications to be enforced on timed Petri nets. Some specifications do not immediately appear to have the necessary form of a predicate on a plant's markings but are still controllable using the techniques of this book. A class of logical specifications on system events and states can be transformed into sets of linear inequalities. The transformation of the Petri net graph to include places that mark the firing of transitions allows for the design of controllers for event based as well as state based constraints. This graph transformation can be combined with the tools for handling uncontrollable transitions to indirectly enforce event based constraints. Finally, the problem of deadlock avoidance can be solved for a large class of Petri nets by enforcing sets of linear inequalities on the plant's state.

Each control place in an invariant based controller represents the slack variable of a single linear inequality on the plant's state. The PN enabling rule with regard to the controller states that if the firing of a transition would cause the marking of any control place to become negative, then that transition is disabled. The result is that the set of all control places enforces the conjunction of the individual constraint inequalities. If the enabling rule is changed so that a transition is only disabled by the controller if its firing would cause all control places to have negative markings, then the controller is enforcing a (generally nonconvex) disjunction of linear inequalities. This extended PN definition greatly expands the kinds of constraints that can be enforced by invariant based controllers while making only a minor modification to the standard Petri net behavioral rules. This technique is particularly important for handling specifications on plants with uncontrollable and unobservable transitions.

Computational efficiency is one of the goals of the supervision techniques presented in this book. An invariant based controller is computed very efficiently by a single matrix multiplication, and its size grows polynomially with the number of specifications. Since the controller is a Petri net, control actions are also simple to compute online. At every iteration, the Petri net enabling inequality is checked for each transition and a linear difference equation is used to update the controller state.

Handling uncontrollable and unobservable transitions does not add any complexity to the online computation of control actions. The increased complexity is encountered only in the initial control design. Computationally tractable techniques have been presented for this process involving the solution of an integer linear program or through the triangularization of an integer matrix through constrained row operations. Other methods in the book involve finding the invariants or siphons of a Petri net. Computationally, both of these problems reduce to finding elements of the kernel of an integer matrix for which established algorithms exist.

Invariant based supervisors are viable models for real time control implementations. The speed and efficiency with which they are computed also makes them appropriate for online control reconfiguration due to sensor or actuator faults, or dynamically modified system specifications.

Glossary

This glossary provides functional definitions of the Petri net and control terms used throughout the book. The *italicized* words in each definition below have their own separate entries in the glossary, however commonly used terms, such as 'Petri net', are not highlighted. This glossary was written to support the text, providing clarification and summary of the most important concepts; for formal definitions, the reader should consult the references.

AC See *asymmetric choice net*.

actuator failure Damage or loss of an actuator, causing a *transition* in the *plant* to become *uncontrollable*.

admissible constraint A *constraint*, $l^T \mu_p \leq b$, on a *plant* with *initial marking* μ_{po}, such that

1. $l^T \mu_{po} \leq b$, and

2. $\forall \mu_p$ reachable from μ_{po} such that $l^T \mu_p \leq b$, μ_p is an *admissible marking*.

If a constraint does not satisfy both of these conditions, then it is *inadmissible*. A constraint is admissible for a plant with *uncontrollable* transitions described by the *incidence matrix* D_{uc} if

$$l^T D_{uc} \leq 0$$

(See Corollary 4.7 of section 4.5.) A constraint is admissible for a plant with *unobservable transitions* described by the incidence matrix D_{uo} if

$$l^T D_{uo} = 0$$

(See Corollary 4.9 of section 4.5.) See also *constraint transformation*, Proposition 4.5 and Corollaries 4.6 and 4.8 of section 4.5, and Propositions 7.4, 7.5 and Corollary 7.2 of section 7.2.1.

admissible marking A *plant marking*, μ_p, such that

1. $l^T \mu_p \leq b$, and

2. For all markings μ'_p *reachable* from μ_p through the firing of uncontrollable transitions, $l^T \mu'_p \leq b$.

If either of these conditions is not met, then the marking is *inadmissible*.

AGV See *automated guided vehicle*.

arc The directed graph element connecting Petri net *places* to *transitions* and transitions to places. See also *inhibitor arc, weight*.

arc weight See *weight*.

asymmetric choice (AC) net A Petri net such that for all pairs of places, p_1 and p_2,

$$\text{if } p_1 \bullet \cap p_2 \bullet \neq \emptyset, \text{ then } p_1 \bullet \subseteq p_2 \bullet \text{ or } p_2 \bullet \subseteq p_1 \bullet$$

See also *free choice net, extended free choice net*.

asynchronous transfer mode (ATM) switch A device used in communications technology to route data cells.

ATM See *asynchronous transfer mode switch*.

automated guided vehicle (AGV) An unmanned conveyance, used in factory automation; normally guided by tracks. See sections 8.2 and 8.3.

available (tokens) Describes those *tokens* in a *timed Petri net* that can be used to *enable transitions*.

blocked Describes a set of *places* with *dead transitions*.

bounded Describes a Petri net where for all reachable markings μ, $\mu \leq b$ for some finite bound b. See also *conservative*.

buffer A holding area for items waiting to be processed.

bullet notation ($\bullet$) Used to represent input or output sets (*presets* and *postsets*) for either *places* or *transitions*. For example, the set of *input places* of the *transition t* is $\bullet t$, while the *output places* are $t \bullet$.

cat and mouse problem A "toy problem" of *discrete event system supervisory* control. The goal is to control the doors of a house such that a cat and a mouse enjoy maximum freedom of movement while never occupying the same room. See section 8.1.

circular wait (CW) See *wait*.

clock A PN structure in a *timed Petri net* that relates to the current value of absolute time, e.g., time of day.

closed loop Describes the *controlled system:* a *plant* and feedback *controller*.

colored Petri net An augmented Petri net model that contains more than one type of *token*. Each token is associated with a "color," and the *transition enabling condition* and state evolution law deal not only with the number of tokens but with their colors as well. Colored Petri nets can be "unfolded" into standard Petri nets.

Commoner's theorem An *extended free choice* Petri net is *live* if and only if every *siphon* in the net contains a *marked trap*. See also *deadlock*, and Propositions 6.2, 6.3, and 6.5 of section 6.2.1.

concurrency Simultaneity. Petri net models are equipped to handle the concurrent *firing* of *transitions* (simultaneous occurrence of events).

conflict Describes the situation of a *place* with more than one *enabled output transition*.

conjunction of constraints A set of linear inequality *constraints*, typically on the *marking* of the *plant*, all of which must be obeyed at all times. Conjunctions are expressed using the vector inequality $L\mu_p \leq b$, which means

$$\bigwedge_{i=1}^{n_c} l_i^T \mu_p \leq b_i$$

where l_i^T is the ith row of L and b_i is the ith element of b. *Invariant based controllers* are synthesized to enforce conjunctions of linear inequalities. See also *disjunctions of constraints*.

conservative Describes a Petri net that is *covered* by a *nonnegative place invariant*. A conservative Petri net is *bounded*, but a bounded net may not be conservative.

constraint A restriction on the allowed behavior of a *plant*. See also *admissible constraint, equality constraint, forbidden state problem, generalized mutual exclusion constraint, logical constraint, mutual exclusion constraint, place constraint, transition constraint*.

constraint transformation Used to create an *admissible constraint* that will not allow any behaviors prohibited by the original inadmissible constraint. Theorem 4.12 of section 4.5 describes constraint transformation and *invariant based controller* synthesis. Given a *plant* with *incidence matrix* D_p, *uncontrollable transitions* described by D_{uc}, *unobservable transitions* described by D_{uo}, and a constraint $L\mu_p \leq b$, assume R_1 and R_2 meet (4.21) and (4.22) with $R_1 + R_2 L \neq 0$ and let

$$\begin{bmatrix} R_1 & R_2 \end{bmatrix} \begin{bmatrix} D_{uc} & D_{uo} & -D_{uo} & \mu_{p_0} \\ LD_{uc} & LD_{uo} & -LD_{uo} & L\mu_{p_0} - b - 1 \end{bmatrix} \leq \begin{bmatrix} 0 & 0 & 0 & -1 \end{bmatrix}$$

Then the controller

$$\begin{aligned} D_c &= -(R_1 + R_2 L)D_p = -L'D_p \\ \mu_{c_0} &= R_2(b+1) - 1 - (R_1 + R_2 L)\mu_{p_0} = b' - L'\mu_{p_0} \end{aligned}$$

exists and causes all subsequent markings of the closed loop system (3.6) to satisfy the constraint $L\mu_p \leq b$ without attempting to inhibit uncontrollable transitions and without detecting unobservable transitions See chapter 5 for computational techniques for transforming constraints and synthesizing controllers.

control invariant Describes a *supervisory* control constraint that has the same closed form independently of the *plant's uncontrollable transitions*. A control invariant constraint is *admissible*.

controlled siphon A *siphon* that remains *marked* for all *reachable markings*. A trap-controlled siphon contains an initially marked *trap*, an invariant-controlled siphon's marking is guaranteed by the presence of a *place invariant*. See section 6.2.1.

controlled system The *closed loop plant/controller* system.

controller See *controller net, invariant based controller, supervisor*.

controller net A Petri net modeled controller, usually employing feedback and enforcing a *supervisory* control law.

cover A nonnegative integer vector x covers a set of *places* S, when $S \subseteq \|x\|$. See *support*.

critical siphon Associated with a circular *wait*, a *siphon* that indicates the set of places that must be emptied for the circular wait to be *blocked*.

critical subsystem A minimal *covering* of the places of a *circular wait* by a *nonnegative place invariant*.

CW (Circular wait.) See *wait*.

dead Describes a *transition* that is currently *disabled* and will never become *enabled* for any possible *reachable marking*. See also *deadlock, live*.

deadlock Condition in which no *transition* in a Petri net is able to fire, i.e., every transition is *dead*. See also *live*.

deadlock avoidance *Supervisory* control policy of restricting *plant* behavior such that *deadlock* can not occur. See also *liveness insurance* and chapter 6.

DEDS Discrete event dynamic system, see *discrete event system*.

defect The quantity $x^T D$, where $x \in \mathbb{Z}^n$, and $D \in \mathbb{Z}^{n \times m}$ is a Petri net *incidence matrix*. The term defect is normally used in discussions of *siphons*, since if $\|x\|$ (see *support*) is a siphon, then $x^T D \leq 0$, while if $x^T D = 0$, then x is a *place invariant*.

DES See *discrete event system*.

direct (enforcement of constraints) See *transition constraints*.

disabled Describes a *transition* that fails to meet the *enabling condition*.

discrete event dynamic system See *discrete event system*.

discrete event system (DES) A dynamic system model with state changes driven by the occurrence of individual events. The state space of a DES is a possibly infinite discrete set. State-to-state transitions are forced by the ordered occurrence of events from a discrete and (almost always) finite set.

disjunction of constraints A set of linear inequality *constraints* on the *marking* of the *plant*, at least one of which must be satisfied at any given time. Disjunctions are expressed

$$\bigvee_{i=1}^{n_c} l_i^T \mu_p \leq b_i$$

A procedure for modifying the *enabling condition* of an *invariant based controller* in order to enforce disjunctions of linear inequalities is described in section 5.4. Disjunctions are important for the realization of *maximally permissive* control laws corresponding to *inadmissible* linear inequality constraints. See *tree structure* and sections 4.4 and 5.3. See also *conjunctions of constraints*.

double integrator A continuous dynamic system with an output that is the double integral of the input.

dummy place A *place* that is temporarily added to the *plant* in order to facilitate the synthesis of an *invariant based controller*.

EFC See *extended free choice net*.

enabled Describes a transition that meets the *enabling condition*.

enabling condition The rule that determines if Petri net *transitions* can *fire*. A given transition is *enabled* if all of its *input places* contain a number of tokens greater than or equal to the *weight* of the *arc* directed from each of these places to the transition. Mathematically, a *firing vector* $q \in \mathbb{Z}^m, q \geq 0$, represents a valid firing if

$$\mu \geq D^- q$$

where μ is the current *state* and D^- is the place-to-transition component of the *incidence matrix*. If the net contains no *self-loops*, the enabling condition can be expressed

$$\mu + Dq \geq 0$$

where D is the complete incidence matrix. Special care must be taken when using this form of the enabling condition when q contains multiple nonzero elements (*concurrent* transition firings).

All transitions that have fired were first enabled, but an enabled transition may not necessarily fire. Rules for determining which among the set of currently enabled transitions will fire are up to the net designer and the needs of the particular application. After choosing a valid firing vector from the set of vectors that meet the enabling condition, the state will evolve according to the rules described in the entry *Petri net*.

equality constraint Constraint of the form $L\mu_p = b$. Attempts to enforce equality constraints with an *invariant based controller* may lead to *deadlock*. See section 7.1.

equivalent constraints Two constraints are equivalent for a given *plant* when the set of behaviors allowed by one constraint is equal to the set allowed by the other.

excess variable A nonnegative scalar used to convert an inequality into an equality: $l^T \mu_p \geq b \Rightarrow l^T \mu_p - \mu_c = b$. See also *slack variable*.

extended free choice (EFC) net A Petri net such that for every *arc* $p \rightarrow t$ there exists an arc from all *input places* of t to all *output transitions* of p. See also *free choice net, asymmetric choice net*.

FC See *free choice net*.

fire The state-changing action of a *transition* in a Petri net. A transition must be *enabled* for it to fire.

firing vector A nonnegative integer vector q with m elements corresponding to the m *transitions* in a Petri net. Often q is constrained such that each element is binary valued, thus the *firing of transition* i corresponds to $q_i = 1$. See also *enabling condition, marking vector*.

firing vector constraint See *transition constraint*.

first order wait See *wait*.

flow relation See *Petri net*.

forbidden state problem A *supervisory* control task of restricting the behavior of a *plant* such that it never enters a given set of (forbidden) states.

formal language See *language*.

free choice (FC) net A Petri net such that for every *arc* from a *place* p to a *transition* t $(p \rightarrow t)$,

1. t is the only output transition of p (no *conflict*), or

2. p is the only input place of t (no *synchronization*).

See also *extended free choice net, asymmetric choice net*.

general circular wait See *wait*.

general tree structure See *tree structure*.

generalized mutual exclusion constraint (GMEC) Constraints on the allowed markings (μ_p) of a plant defined by the set

$$\mathcal{M}(l, b) = \{\mu_p \in \mathbb{Z}^n, \mu_p \geq 0 | l^T \mu_p \leq b\}$$

GMEC's can be used to realize both serial and parallel *mutual exclusion constraints*. GMEC's can be enforced using *monitors* or *invariant based controllers*.

generating family A set of *siphons* (or *traps*) that covers the set of all a net's siphons (or traps). All siphons (traps) in a Petri net can be expressed as unions of members of the generating family of siphons (traps).

GMEC See *generalized mutual exclusion constraint*.

graph transformation (for handling *transition constraints*) A transformation of each *transition* indicated in a constraint into two transitions with a *place* in between. The transition constraint is then rewritten as a *place constraint* in terms of the *dummy places*. See section 7.2.

hybrid Describes a system with mixed continuous and discrete dynamics.

inadmissible constraint See *admissible constraint*.

inadmissible marking See *admissible marking*.

incidence matrix Description of the *weighted* flow relation between the Petri net *places* and *transitions*. The incidence matrix $D \in \mathbb{Z}^{n \times m}$, where n is the number of places and m is number of transitions, is composed of two other matrices:

$$D = D^+ - D^-$$

where $D^+, D^- \geq 0$. D^+ contains the weights of *arcs* from *transitions* to *places*, thus if transition j contains an arc with weight w to place i, then $D_{ij}^+ = w$. Similarly, D^- contains the weights of arcs from places to transitions. If the Petri net contains no *self-loops*, then it is not necessary to maintain separate D^+ and D^- matrices, since all the information will be contained in D. The incidence matrix is fundamental to the evolutionary rules of a Petri net; see *Petri net* and *enabling condition*. See also *place invariant, transition invariant*.

indirect (enforcement of constraints) See *transition constraints*.

inhibitor arc An *arc* $p \rightarrow t$ that prevents *transition* t from firing when *place* p contains a number of *tokens* greater than or equal to the arc *weight*. Standard ordinary *Petri nets* do not contain inhibitor arcs.

initial marking Initial state, initial condition. The distribution of *tokens* throughout the *places* of a Petri net before any *transitions* have *fired*. See also *marking, marking vector*.

input place See *place*.

input transition See *transition*.

invariant See *place invariant, transition invariant*.

invariant based controller A *supervisory* controller for enforcing linear inequalities on the *reachable markings* of a *plant* Petri net. Given a plant with *incidence matrix* $D_p \in \mathbb{Z}^{n \times m}$ and *initial marking* μ_{p_0}, a supervisor for enforcing $L\mu_p \leq b$, $L \in \mathbb{Z}^{n_c \times n}$ exists iff

$$b - L\mu_{p_0} \geq 0$$

The incidence matrix and initial marking of the invariant based controller are given by

$$D_c = -LD_p$$
$$\mu_{c_0} = b - L\mu_{p_0}$$

and the closed loop system is

$$D = \begin{bmatrix} D_p \\ D_c \end{bmatrix} \qquad \mu_0 = \begin{bmatrix} \mu_{p_0} \\ \mu_{c_0} \end{bmatrix}$$

See Theorem 3.2 in section 3.2 for more details.

The controller described above will enforce the *conjunction* of a set of linear inequalities on the marking of a plant by making them *place invariants* of the closed loop system. Section 5.4 describes how a modification to the controller's *enabling condition* can be used to create invariant based controllers that enforce *disjunctions of linear inequalities*.

If the plant contains any *uncontrollable* or *unobservable transitions*, then the constraints enforced by an invariant based controller must be *admissible*. See also *constraint transformation*.

invariant-controlled siphon See *controlled siphon*.

kernel A set of vectors corresponding to a matrix D, either $\{x|x^T D = 0\}$ or $\{y|Dy = 0\}$. The "left hand" or "right hand" meaning is normally left up to context. Petri net *place invariants* are elements of the left hand (x form) kernel of the *incidence matrix* D, and *transition invariants* are elements of the right hand (y form) kernel of D. A kernel is often represented with a matrix, the rows (or columns) of which form a linearly independent basis for the kernel.

language The languages accepted by automata are categorized as follows.

type 3 Regular languages are accepted by finite state machines.

type 2 Context-free languages are accepted by pushdown automata.

type 1 Context-sensitive languages are accepted by linear bounded automata and bounded two-pushdown automata.

type 0 The most complex of the formal languages, phrase structure languages are accepted by Turing machines and two-pushdown automata.

Each language type i is a subset of language type $i - 1$. The set of *Petri net* languages includes all regular languages, intersects the context-free languages, and is a subset of the context-sensitive languages.

least restrictive See *maximally permissive*.

live Describes a *transition* for which there exists some sequence of valid *firings* such that it is *enabled*, no matter what *reachable marking* the Petri net occupies. A live transition may be fired again and again no matter how the state evolves.

A Petri net is called live when all of its transitions are live. A live net is guaranteed to be *deadlock*-free, however, a deadlock-free net may not necessarily be completely live. See sections 6.2.1 and 6.2.2.

liveness insurance *Supervisory* control policy of restricting the behavior of a non-*live* Petri net such that all of its *transitions* remain live. See also *deadlock avoidance* and chapter 6.

logical constraint A constraint on *plant* behavior expressed with boolean variables and logical operators (AND, OR, NOT, etc.).

M-1 A robot used in manufacturing.

marked graph (MG) A Petri net in which all places have a single input and a single output transition. Also called a T-system.

marking The distribution of *tokens* throughout the places of a Petri net. The marking uniquely identifies the *state* of the net. See also *marking vector, initial marking.*

A place, or a set of places, is called *marked* when any of the places contains a token. A set of empty places is called *unmarked.*

marking vector A nonnegative integer vector μ with n elements corresponding to the n places in a Petri net. The value of the ith element of μ is equal to the number of *tokens* held in place i, i.e., the *marking* of place i. See also *initial marking, firing vector.*

marking vector constraint See *place constraint.*

maximally permissive Describes a *supervisor* that gives a *plant* the highest possible amount of freedom during its evolution. A maximally permissive supervisor will only disable an event in the evolution of a plant when the occurrence of that event would either violate a constraint directly or indirectly through the *firing* of *uncontrollable transitions.*

MG See *marked graph.*

minimally restrictive See *maximally permissive.*

monitor A Petri net controller developed by [Giua et al., 1992] for enforcing *generalized mutual exclusion constraints* on the reachable markings of *plant* Petri nets. Monitors are constructed like *invariant based controllers.*

mutual exclusion constraint A restriction on the weighted sum of *tokens* allowed in a set of *places*, usually involving the use of *finite resources.* A *parallel mutual exclusion* indicates that only one process in a set of processes is allowed to use a given resource at any one time. A *sequential mutual exclusion* is similar, but the processes will receive the necessary resource following a set order. See also *generalized mutual exclusion constraint.*

nonconvex control law A *supervisory* control specification that allows nonconvex sets of *markings.* $L\mu_p \leq b$ always describes a convex set of states, while *disjunctions* of inequalities are, in general, nonconvex.

nonnegative place invariant A *place invariant* $x \geq 0$.

OPC See *output-port controller.*

optimal control See *maximally permissive.*

output place See *place.*

output-port controller (OPC) Section of an *asynchronous transfer mode switch* used to store and then transmit data cells. See section 8.5.

output transition See *transition.*

permissive See *maximally permissive.*

Petri net (PN) A directed bipartite graph capable of modeling *discrete event systems* with *concurrent* events. The structure of a Petri net is described by (P, T, D^+, D^-) where P and T are disjoint sets representing the vertices of the graph, known as *places* and *transitions*, and D^+ and D^- are integer matrices with nonnegative elements representing the flow relation between the two vertex types. An *arc* with *weight* D_{ij}^+ from transition j to place i indicates that when transition j fires, place i will receive D_{ij}^+ *tokens*. An arc with weight D_{kj}^- from place k to transition j indicates that place k must contain at least D_{kj}^- tokens before transition j is allowed to fire and that when transition j fires, place k will lose D_{kj}^- tokens.

The *state* or *marking vector*, μ, of a Petri net is determined by the distribution of tokens throughout the net's places; μ_i is the (nonnegative) number of tokens in place p_i. The state dynamically evolves according to

$$\begin{aligned} \mu(0) &= \mu_0 \\ \mu(k+1) &= \mu(k) + Dq(k) \end{aligned}$$

$$D \in \mathbb{Z}^{n \times m}, \mu \in \mathbb{Z}^n, q \in \mathbb{Z}^m, (\mu, q \geq 0)$$

where μ_0 is the *initial marking*, q is the *firing vector*, D is the *incidence matrix*, and k is an iteration counter.

If the Petri net contains *self-loops*, then additional information is needed to completely define the allowed evolution of the net; see *enabling condition*.

See also *asymmetric choice net, extended free choice net, free choice net, marked graph, state machine, colored Petri net, timed Petri net, bounded, conservative, safe, inhibitor arc*.

P-invariant See .

piston rod robotic assembly cell A manufacturing application in which robots install a piston rod in an engine block. See section 8.4.

pivot The element in a matrix that is the current focus of an algorithm that performs column or *row operations*.

place One of the two distinct types of Petri net vertices (the other type is the *transition*). A place is a storage cell for *tokens*. The distribution of tokens among the places determines the state or *marking* of the net.

The *input places* of a *transition*, $\bullet t$, (or set of transitions, $\bullet T$) form the set of places with *arcs* directed to that transition (or to any transition in the set). The *output places* of a transition, $t\bullet$, (or set of transitions, $T\bullet$) form the set of places receiving *arcs* from that transition (or from any transition in the set). See also *preset, postset*.

place constraint A constraint placed on the *reachable markings* of a *plant*, usually expressed as a linear inequality, $l^T \mu_p \leq b$, or $L\mu_p \leq b$. See also *generalized mutual exclusion constraint, transition constraint*.

place invariant An integer vector corresponding to a weighted sum of *tokens* in a set of *places* that remains constant for all *reachable markings*. A place invariant is defined as any $x \in \mathbb{Z}^n$ that satisfies

$$x^T \mu = x^T \mu_0$$

where μ_0 is the net's *initial marking*, and μ represents any subsequent marking. Place invariants can be computed by finding integer solutions to

$$x^T D = 0$$

where $D \in \mathbb{Z}^{n \times m}$ is the *incidence matrix* of the Petri net, thus place invariants are elements of the (left hand) *kernel* of D. See also *nonnegative place invariant, transition invariant.*

plant   Any system that is to be controlled.

PN   See *Petri net.*

positive row operation   See *row operation.*

postset   The set of *places* (*transitions*) that receive *arcs* from a given transition or set of transitions (place or set of places). See also *preset.*

PPN   See *production Petri net.*

premultiply   To multiply a matrix from the left. A premultiplies B in the expression AB.

preset   The set of *places* (*transitions*) that direct *arcs* to a given transition or any of a set of transitions (place or any of a set of places). See also *postset.*

process net   A Petri net modeled *plant.*

production Petri net (PPN)   A Petri net framework for modeling manufacturing systems. A PPN includes strings of *places*, representing steps in one or more processes, with *resource* usage constraints imposed on them.

reachable   Describes any *marking* that may be attained during the legal evolution of the Petri net state. The *initial marking* and *enabling condition* (which incorporates the structure of the net) determine the set of reachable markings.

redundant   Describes constraints on the behavior of a plant that are naturally satisfied without *supervision*. Redundant *place constraints* are *admissible* (see Corollary 4.8 of section 4.5).

regulation circuit   A pair of *places* connected to a Petri net to insure that two sequences of events occur alternately in order. See section 6.2.2.

resource   An item needed to perform some task or step in a process. Certain *places* in a Petri net may represent processes that require the use of one or more finite resources, thus constraints are placed on the number of *tokens* that can occupy these places of the net.

row operation   Transformation of a row in a matrix. The elementary row operations on an integer matrix are

1. Interchange any two rows.

2. Multiply any row by a nonzero integer.

3. Add to any row another row multiplied by a nonzero integer.

Column operations are defined similarly. The term 'positive row operation' indicates that the multiplying integer of operation 2 or 3 is positive.

S-380   A robot used in manufacturing.

S-invariant   See *place invariant.*

S-system   See *state machine.*

safe   Describes a Petri net that can never contain more than one *token* in any given *place*. The *marking* of a safe net is binary-valued.

safety constraint   A constraint restricting the *plant* from occupying certain *forbidden states.*

self-loop Describes the situation of a *transition* with input and output *arcs* involving the same *place*. Corresponding elements in the D^+ and D^- components of the *incidence matrix* will both have positive values when a self-loop is present, thus some information about the true flow relation of the net may not be present in D. See also *enabling condition*.

sensor failure Damage or loss of a sensor, causing a *transition* in the *plant* to become *unobservable*.

simple circular wait See *wait*.

simple (constraints) Constraint $l_1^T \mu_p \le b_1$ is simpler than $l_2^T \mu_p \le b_2$ if $l_1 < l_2$.

sinks The *postset* of a *place* or *transition* (or group of places or transitions). See also *sources*.

siphon A set of *places* S where

$$\bullet S \subseteq S \bullet$$

S is a *minimal siphon* iff there does not exist another siphon P such that $P \subset S$. A siphon that loses all of its *tokens* will remain empty for all future *reachable markings*. See also *Commoner's theorem, controlled siphon, critical siphon, generating family, trap*.

slack variable A nonnegative scalar used to convert an inequality into an equality: $l^T \mu_p \le b \Rightarrow l^T \mu_p + \mu_c = b$. The *places* of an *invariant based controller* play the part of slack variables. The value of a slack variable is called the *slack*. See also *excess variable*.

SM See *state machine*.

sources The *preset* of a *place* or *transition* (or group of places or transitions). See also *sinks*.

state (of Petri net) See *marking*.

state feedback See *static state feedback*.

state machine (SM) A Petri net in which all transitions have one input and one output place. Also called an *S-system*.

state vector See *marking vector*.

static state feedback A control law that corresponds to a direct mapping from the current *plant* state (or *marking*) to sets of *enabled* and *disabled transitions*. *Invariant based controllers* enforce static state feedback laws (see Proposition 7.6 of section 7.5).

stellen German word (used by Petri) for *place*. The letter 'S' is sometimes used to designate sets, vectors, or other items associated with PN places.

structural properties Properties that depend only on the topological structure of the Petri net (the flow relation) and not on the net's initial marking. Examples: *place* and *transition invariants*.

supervisor A controller that leaves the *plant* free to evolve, only acting to prohibit certain undesirable behaviors. *Invariant based controllers* are supervisors.

support Let $x \in \mathbb{Z}^n$ be an integer vector corresponding to the *places* of a Petri net. The support of x, $\|x\|$, is the set of places corresponding to the nonzero entries in x. See also *cover*.

synchronization Situation in which a *transition* has more than one *input place*.

T-invariant See *transition invariant*.

T-system See *marked graph*.

three tanks problem A process control example involving three fluid-filled tanks with drains and a pump that can add fluid to each of them. The goal is to control the operation of the pump such that the fluid level of each tank stays within desirable bounds. See section 8.6.

timed Petri net A Petri net augmented with real time information associated with its state evolution. Timed evolution can be modeled as the amount of time required to *fire* the *transitions*, or the amount of time that a *token* must reside in a *place* before it can be used to *enable* a transition. See also *available, waiting* and section 7.4.

timer A PN structure in a *timed Petri net* that records elapsed time since the occurrence of some event.

token Counters used to indicate the *marking* of Petri net places. The marking of a place is equal to the number of tokens held in that place.

transformation See *constraint transformation, graph transformation*

transition One of the two distinct types of Petri net vertices (the other type is the *place*). The *firing* of transitions is the vehicle by which the *marking* of places is changed, i.e., the firing of a transition is a state changing event.

The *input transitions* of a place, •*p* (or set of places, •*P*) form the set of transitions with *arcs* directed to that place (or to any place in the set). The *output transitions* of a place, *p*•, (or set of places, *P*•) form the set of transitions receiving arcs from that place (or any place in the set). See also *preset, postset*.

transition constraint A constraint placed on the *transition firings* of the *plant*, usually expressed as a linear inequality $f^T q \leq b$. A mixed *place*/transition constraint has the form $l^T \mu_p + f^T q \leq b$.

In the *direct* enforcement of a transition constraint, the *supervisor* actively *enables* and *disables* the transitions described by the constraint in order to make it true. In the *indirect* enforcement of a transition constraint, the supervisor will forbid those states that leave transitions enabled such that if they were to *fire* the constraint would be violated. See section 7.2. See also *graph transformation, place constraint*.

transition invariant Integer vectors that represent a set of *transition firings* that will cause the *marking* of a net to cycle, leaving it in the state that it held before the start of the cycle. If

$$
\begin{aligned}
\mu(0) &= \mu_0 \\
\mu(1) &= \mu_0 + Dq(0) \\
\mu(2) &= \mu(1) + Dq(1) \\
&= \mu_0 + Dq(0) + Dq(1) \\
&\vdots \\
\mu(N) &= \mu_0 + D(q(0) + \cdots + q(N-1)) = \mu_0
\end{aligned}
$$

where μ is the net's *marking vector*, D is the *incidence matrix*, and q is the *firing vector*, then $y = \sum_{i=0}^{N-1} q(i)$ is a transition invariant. Transition invariants lie in the (right hand) *kernel* of D:

$$Dy = 0$$

The existence of a transition invariant does not imply that it is actually possible to *fire* the indicated transitions; the *initial conditions* of a net may prohibit it. Instead a transition invariant indicates that if it is possible to fire the given set of transitions, in any order, the state of the net will return to its initial condition at the end of the sequence. See also *place invariant.*

trap A set of *places* S where

$$S\bullet \subseteq \bullet S$$

S is a *minimal trap* iff there does not exist another trap P such that $P \subset S$. Once a trap becomes *marked*, it will remain marked for all future reachable markings. See also *generating family, siphon.*

trap-controlled siphon See *controlled siphon.*

tree structure (in Petri nets) Describes a Petri net (or *VDES*) that has no loops, its *places* have no more than one output *transition*, and its graph is such that places on one level of the graph act as the exclusive set of *sources* for the transitions on the next level, which then act as the sources for the places on the next level, etc. There are two subcategories of the general tree structure.

type 1 Every transition has at most one output *arc.*

type 2 Every transition has at most one input arc.

If the *uncontrollable portion* of a *plant* has a type 1 tree structure, then the *maximally permissive admissible* control law for a linear inequality constraint will have the form of a disjunction of finitely many linear inequalities. If the uncontrollable portion of a plant has a type 2 tree structure, then the maximally permissive admissible control law for a linear inequality constraint will also have the form of a single linear inequality [Li and Wonham, 1994].

type See *language* or *tree structure.*

uncontrollable transition A *plant transition*, the *firing* of which can not be inhibited by an external action. See also *admissible constraint, unobservable transition.*

uncoupled Describes a mixed *place/transition constraint* $l^T \mu_p + f^T q \leq b$, $f \geq 0$ where

$$T_l \cap T_f = \emptyset$$

where T_l is the set of transitions that are connected to the controller induced by the $l^T \mu_p$ portion of the constraint, and T_f is the set of transitions connected to the controller induced by the $f^T q$ portion of the constraint.

unobservable transition A *plant transition*, the firings of which can not be directly detected or measured. For a Petri net modeled controller, the ability of the controller to inhibit a transition is intrinsically tied with its ability to change state based on observations of that transition, thus unobservable transitions are also *uncontrollable* for Petri net controllers. See also *admissible constraint.*

unreliable machine A model of a *plant* that includes details of possible break-down or malfunction. See section 8.3.

unweighted constraint A linear *place constraint* $l^T \mu_p \leq b$ such that l is composed only of 1's and 0's.

VDES See *vector discrete event system.*

vector discrete event system (VDES) A *DES* model that is equivalent in one of its forms to a Petri net. [Li and Wonham, 1993, Li and Wonham, 1994, Li and Wonham, 1995].

wait A relationship between finite *resources*. If the release of resource r_i is conditional on the availability of r_j, then r_i waits for r_j, or $r_i \hookrightarrow r_j$. If r_i can be released immediately upon the availability of r_j, then there is a *first order wait* relation between r_i and r_j: $r_i \rightarrow r_j$.

A *circular wait (CW)* is a set of resources that all wait on each other. A *simple circular wait* is described as $r_1 \rightarrow r_2 \rightarrow \ldots \rightarrow r_q \rightarrow r_1$. A set of resources C is a *general circular wait* if $\forall r_i, r_j \in C, r_i \hookrightarrow r_j$. See also *critical siphon, critical subsystem*.

waiting (tokens) Describes those *tokens* in a *timed Petri net* that can not yet be used to *enable transitions*.

weight A nonnegative integer associated with an *arc*. When the arc is from a *transition* to a *place*, the weight indicates the number of tokens that will be transfered to the place upon the *firing* of the transition. When the arc is from a place to a transition, the weight indicates the minimum number of tokens that must be in the place for the transition to be *enabled* and the number of tokens that will be removed from the place if the transition fires.

well-marked Describes a Petri net in which all of its *siphons* are initially *marked*, i.e., each siphon starts with at least one *token*.

References

Antsaklis, P. J. and Kantor, J. C. (1995). Intelligent control for high autonomy process control systems. Technical Report of the ISIS Group ISIS-95-004, University of Notre Dame, Notre Dame, IN. Presented at ISPE '95, Snowmass Village, Colorado, July, 1995.

Antsaklis, P. J., Lemmon, M. D., and Stiver, J. A. (1993a). Learning to be autonomous: Intelligent supervisory control. Technical Report of the ISIS Group ISIS-93-003, University of Notre Dame, Notre Dame, IN.

Antsaklis, P. J. and Michel, A. N. (1997). *Linear Systems*. McGraw-Hill.

Antsaklis, P. J. and Passino, K. M., editors (1993). *An Introduction to Intelligent and Autonomous Control*. Kluwer Academic Publishers.

Antsaklis, P. J., Stiver, J. A., and Lemmon, M. D. (1993b). Hybrid system modeling and autonomous control systems. In Grossman, R. L., Nerode, A., Ravn, A. P., and Rischel, H., editors, *Hybrid Systems*, volume 736 of *Lecture Notes in Computer Science*, pages 366–392. Springer-Verlag, Berlin; New York.

Avrunin, G. S., Buy, U. A., and Corbett, J. C. (1991). Integer programming in the analysis of concurrent systems. In Larsen, K. G. and Skou, A., editors, *Computer Aided Verification, Third International Workshop Proceedings*, volume 575 of *Lecture Notes in Computer Science*, pages 92–102. Springer-Verlag.

Banaszak, Z. A. and Krogh, B. H. (1990). Deadlock avoidance in flexible manufacturing systems with concurrently competing process flows. *IEEE Transactions on Robotics and Automation*, 6(6):724–734.

Barkaoui, K. (1995). Liveness of Petri nets and its relations with deadlocks, traps, and invariants. Report 92-06, Laboratoire CEDRIC-CNAM, Paris, France.

Barkaoui, K. and Abdallah, I. B. (1995). Deadlock avoidance in FMS based on structural theory of petri nets. In *IEEE Symposium on Emerging Technologies and Factory Automation*, volume 2, pages 499–510, Piscataway, NJ. IEEE.

Bogdan, S. and Lewis, F. L. (1997). Matrix approach to deadlock avoidance of dispatching in multi-class finite buffer reentrant flow lines. In *Proceedings of the 1997 IEEE International Symposium on Intelligent Control*, pages 397–402, Piscataway, NJ. IEEE.

Boissel, O. (1993). Optimal feedback control design for discrete-event process systems using simulated annealing. Master's thesis, University of Notre Dame, Notre Dame, IN.

Boucher, T. O. and Jafari, M. A. (1992). Design of a factory floor sequence controller from a high level system specification. *Journal of Manufacturing Systems*, 11(6):401–417.

Brand, K.-P. and Kopainsky, J. (1988). Principles and engineering of process control with petri nets. *IEEE Transactions on Automatic Control*, 33(2):138–149.

Carroll, J. (1989). *Theory of Finite Automata*. Prentice Hall, Engelwood Cliffs, NJ.

Cassandras, C. G. (1993). *Discrete Event Systems*. Richard D. Irwin, Inc. and Aksen Associates, Inc.

Chase, C., Serrano, J., and Ramadge, P. J. (1993). Periodicity and chaos from switched flow systems: Contrasting examples of discretely controlled continuous systems. *IEEE Transactions on Automatic Control*, 38(1):70–83.

Cohen, G., Dubois, D., Quadrat, J. P., and Viot, M. (1985). A linear-system theoretic view of discrete event processes and its view for performance evaluation in manufacturing. *IEEE Transactions on Automatic Control*, 30(3):210–220.

Corbett, J. C. and Avrunin, G. S. (1995). Using integer programming to verify general safety and liveness properties. *Formal Methods in System Design: An International Journal*, 6(1):97–123.

David, R. and Alla, H. (1987). Continuous Petri nets. In *Proceedings of the 8th European Workshop on Application and Theory of Petri Nets*, Zaragosa.

Desel, J. and Esparza, J. (1995). *Free Choice Petri Nets*. Cambridge University Press.

Desrochers, A. A. and Al-Jaar, R. Y. (1995). *Applications of Petri Nets in Manufacturing Systems*. IEEE Press, Piscataway, NJ.

Ezpeleta, J., Colom, J. M., and Martínez, J. (1995). A Petri net based deadlock prevention policy for flexible manufacturing systems. *IEEE Transactions on Robotics and Automation*, 11(2):173–184.

Ezpeleta, J., Couvreur, J. M., and Silva, M. (1993). A new technique for finding a generating family of siphons, traps, and ST-components. applications to colored Petri nets. In Rozenberg, G., editor, *Advances in Petri Nets*, volume 674 of *Lecture Notes in Computer Science*, pages 126–147. Springer-Verlag, Berlin; New York.

Fang, S.-C. and Puthenpura, S. (1993). *Linear Optimization and Extensions: Theory and Algorithms*. Prentice Hall, Engelwood Cliffs, NJ.

Giua, A. and DiCesare, F. (1994). Blocking and controllability of Petri nets in supervisory control. *IEEE Transactions on Automatic Control*, 39(4):818–823.

Giua, A., DiCesare, F., and Silva, M. (1992). Generalized mutual exclusion constraints on nets with uncontrollable transitions. In *Proceedings of the 1992 IEEE International Conference on Systems, Man, and Cybernetics*, pages 974–979, Chicago, IL.

Guo, D. L., DiCesare, F., and Zhou, M. C. (1993). A moment generating function based approach for evaluating extended stochastic petri nets. *IEEE Transactions on Automatic Control*, 38(2):321–327.

Hervé, P. H. and Proth, J.-M. (1989). Performance evaluation of job-shop systems using timed event graphs. *IEEE Transactions on Automatic Control*, 34(1):3–9.

Ho, L., editor (1992). *Discrete Event Dynamic Systems: Analyzing Complexity and Performance in the Modern World*. IEEE Press, NY.

Ho, Y.-C. and Cao, X.-R. (1991). *Perturbation Analysis of Discrete Event Dynamic Systems*. Kluwer Academic Publishers, Norwell, MA.

Holloway, L. E. and Krogh, B. H. (1990). Synthesis of feedback logic for a class of controlled Petri nets. *IEEE Transactions on Automatic Control*, 35(5):514–523.

Holloway, L. E. and Krogh, B. H. (1994). Controlled Petri nets: A tutorial survey. In Cohen, G. and Quadrat, J.-P., editors, *Lecture Notes in Computer Science*, volume 199, pages 158–168. Springer-Verlag, Berlin; New York. Eleventh International Conference on Analysis and Control, Discrete Event Systems.

Holloway, L. E., Krogh, B. H., and Giua, A. (1997). A survey of Petri net methods for controlled discrete event systems. *Discrete Event Dynamic Systems: Theory and Applications*, 7(2):151–190.

Hooker, J. N. (1988). A qualitative approach to logical inference. *Decision Support Systems*, (4):45–69.

Huang, H.-H., Lewis, F. L., Pastravanu, O. C., and Gurel, A. (1995). Flow-shop scheduling design in an FMS matrix framework. *Control Engineering Practice*, 3(4):561–568.

Huang, H.-H., Lewis, F. L., and Tacconi, D. A. (1996). Deadlock analysis using a new matrix-based controller for reentrant flow line design. In *IECON Proceedings (Industrial Electronics Conference)*, volume 1, pages 463–468. IEEE, Los Alamitos, CA.

Ichikawa, A. and Hiraishi, K. (1988). Analysis and control of discrete event systems represented by Petri nets. In Varaiya, P. and Kurzhanski, A. B., editors, *Discrete Event Systems: Models and Applications*, Lecture Notes in Computer Science. Springer-Verlag, Berlin; New York.

Jensen, K. (1992). *Coloured Petri Nets Basic Concepts, Analysis Methods and Practical Use*, volume 1. Springer-Verlag, Berlin; New York.

Kumar, R. and Garg, V. K. (1995). *Modeling and Control of Logical Discrete Event Systems*. Kluwer Academic Publishers, Norwell, MA.

Labinaz, G., Rudie, K., Ricker, S. L., Sarkar, N., and Bayoumi, M. M. (1997). A hybrid system investigation of fluid-filled tanks. *Submitted to the IEEE Transactions on Automatic Control Special Issue on Hybrid Systems*.

Laftit, S., Proth, J.-M., and Xie, S.-L. (1992). Optimization of invariant criteria for event graphs. *IEEE Transactions on Automatic Control*, 37(5):547–555.

Lautenbach, K. (1987). Linear algebraic techniques for place/transition nets. In *Lecture Notes in Computer Science: Advances in Petri Nets Part I – Petri Nets: Central Models and Their Properties*, volume 254, pages 142–167. Springer-Verlag, Berlin; New York.

Lautenbach, K. and Ridder, H. (1994). Liveness in bounded Petri nets which are covered by T-invariants. In Valette, R., editor, *Application and Theory of Petri Nets*, Lecture Notes in Computer Science, pages 358–375. Springer-Verlag, Berlin; New York.

Lemmon, M. D., Stiver, J. A., and Antsaklis, P. J. (1993). Event identification and intelligent hybrid control. In Grossman, R. L., Nerode, A., Ravn, A. P., and Rischel,

H., editors, *Hybrid Systems*, volume 736 of *Lecture Notes in Computer Science*, pages 265–296. Springer-Verlag.

Lewis, F. L., Huang, H., Tacconi, D., Gürel, A., and Pastravanu, O. (1995). Analysis of deadlocks and circular waits using a matrix model for discrete event systems. Technical report, Automation and Robotics Research Institute, The University of Texas at Arlington, Ft. Worth, TX.

Li, Y. and Wonham, W. M. (1993). Control of vector discrete event systems I – the base model. *IEEE Transactions on Automatic Control*, 38(8):1214–1227. Correction in IEEE TAC v. 39 no. 8, pg. 1771, Aug. 1994.

Li, Y. and Wonham, W. M. (1994). Control of vector discrete event systems II – controller synthesis. *IEEE Transactions on Automatic Control*, 39(3):512–530.

Li, Y. and Wonham, W. M. (1995). Concurrent vector discrete-event systems. *IEEE Transactions on Automatic Control*, 40(4):628–638.

Marsan, M. A., Conte, G., and Balbo, G. (1984). A class of generalized stochastic Petri nets for the performance evaluation of multiprocessor systems. *ACM Transactions on Computer Systems*, 2(2).

Martinez, J. and Silva, M. (1980). A simple and fast algorithm to obtain all invariants of a generalized Petri net. In *Advances in Petri Nets*, number 52 in Lecture Notes in Computer Science. Springer-Verlag, Berlin.

Michel, A. N. and Herget, C. J. (1993). *Applied Algebra and Functional Analysis*. Dover Pulications, Inc., New York.

Minoura, T. and Ding, C. (1991). A deadlock prevention method for a sequence controller for manufacturing control. *International Journal of Robotics and Automation*, 6(3).

Molloy, M. K. (1981). *On the Integration of Delay and Throughput Measures in Distributed Processing Models*. PhD thesis, University of California, Los Angeles, CA.

Moody, J. O. and Antsaklis, P. J. (1995). Petri net supervisors for DES in the presence of uncontrollable and unobservable transitions. In *Proceedings of the 33rd Annual Allerton Conference on Communication, Control, and Computing*, pages 176–185, Monticello, IL.

Moody, J. O. and Antsaklis, P. J. (1996a). Supervisory control of Petri nets with uncontrollable/unobservable transitions. In *Proceedings of the 35th IEEE Conference on Decision and Control*, pages 4433–4438, Kobe, Japan.

Moody, J. O. and Antsaklis, P. J. (1996b). Supervisory control of Petri nets with uncontrollable/unobservable transitions. Technical Report of the ISIS Group ISIS-96-004, University of Notre Dame, Notre Dame, IN.

Moody, J. O. and Antsaklis, P. J. (1997a). Characterization of feasible controls for Petri nets with unobservable transitions. In *Proceedings of the 1997 American Control Conference*, volume 4, pages 2354–2358, Albuquerque, New Mexico.

Moody, J. O. and Antsaklis, P. J. (1997b). Supervisory control using computationally efficient linear techniques: A tutorial introduction. In *Proceedings of 5th IEEE Mediterranean Conference on Control and Systems*, volume Session MP1, Paphos, Cyprus.

Moody, J. O., Antsaklis, P. J., and Lemmon, M. D. (1995a). Automated design of a Petri net feedback controller for a robotic assembly cell. In *Proceedings of 1995 INRIA/IEEE Symposium on Emerging Technologies and Factory Automation*, volume 2, pages 117–128, Paris, France.

Moody, J. O., Antsaklis, P. J., and Lemmon, M. D. (1995b). Feedback Petri net control design in the presence of uncontrollable transitions. In *Proceedings of the 34th IEEE Conference on Decision and Control*, volume 1, pages 905–906, New Orleans, LA.

Moody, J. O., Antsaklis, P. J., and Lemmon, M. D. (1996). Petri net feedback controller design for a manufacturing system. In *Proceedings of the IFAC 13th Triennial World Congress*, volume B, pages 67–72, San Francisco, CA.

Moody, J. O., Yamalidou, K., Lemmon, M. D., and Antsaklis, P. J. (1994). Feedback control of Petri nets based on place invariants. In *Proceedings of the 33rd IEEE Conference on Decision and Control*, volume 3, pages 3104–3109, Lake Buena Vista, FL.

Murata, T. (1989). Petri nets: Properties, analysis, and applications. *Proceedings of the IEEE*, 77(4):541–580.

Murata, T., Komoda, N., Matsumoto, K., and Haruna, K. (1986). A Petri net-based controller for flexible and maintainable sequence control and its applications in factory automation. *IEEE Transactions on Industrial Electronics*, 33(1):1–8.

Passino, K., Michel, A., and Antsaklis, P. J. (1994). Lyapunov stability of a class of discrete event systems. *IEEE Transactions on Automatic Control*, 39(2):269–279.

Passino, K. M. (1989). *Analysis and Synthesis of Discrete Event Regulator Systems*. PhD thesis, Department of Electrical and Computer Engineering, University of Notre Dame, Notre Dame, IN.

Peterson, J. L. (1981). *Petri Net Theory and the Modeling of Systems*. Prentice Hall, Engelwood Cliffs, NJ.

Proth, J.-M. and Xie, X.-L. (1994). Cycle time of stochastic event graphs: Evaluation and marking optimization. *IEEE Transactions on Automatic Control*, 39(7):1482–1486.

Ramadge, P. J. G. and Wonham, W. M. (1989). The control of discrete event systems. *Proceedings of the IEEE*, 77(1):81–97.

Ramamoorthy, C. V. and Ho, G. S. (1980). Performance evaluation of asynchronous concurrent systems using Petri nets. *IEEE Transactions on Software Engineering*, 6(5).

Reisig, W. (1985). *Petri Nets*. Springer-Verlag, Berlin; New York.

Reisig, W. (1992). *A Primer in Petri Net Design*. Springer-Verlag, Berlin; New York.

Révész, G. E. (1983). *Introduction to Formal Languages*. McGraw-Hill.

Ross, K. W. (1986). Optimal dynamic routing in Markov queuing networks. *Automatica*, 22(3):367–370.

Sifakis, J. (1977). Use of Petri nets for performance evaluation. In Beilner, H. and Gelenbe, E., editors, *Measuring, Modelling, and Evaluating Computer Systems*. North Holland Pub., Amsterdam.

Sifakis, J. (1979). Performance evaluation of systems using Petri nets. In Brauer, W., editor, *Advances in Petri Nets*, number 84 in Lecture Notes in Computer Science. Springer-Verlag, Berlin; New York.

Sreenivas, R. S. (1997a). On commoner's liveness theorem and supervisory policies that enforce liveness in free-choice Petri nets. *Systems & Control Letters*, 31(1):41–48.

Sreenivas, R. S. (1997b). On the existence of supervisory policies that enforce liveness in discrete-event dynanic systems modeled by controlled Petri nets. *IEEE Transactions on Automatic Control*, 42(7):928–945.

Sreenivas, R. S. and Krogh, B. H. (1992). On petri net models of infinite state supervisors. *IEEE Transactions on Automatic Control*, 37(2):274–276.

Stiver, J. A. (1995). *Analysis and Design of Hybrid Control Systems*. PhD thesis, Department of Electrical Engineering, University of Notre Dame, Notre Dame, IN.

Stiver, J. A., Antsaklis, P. J., and Lemmon, M. D. (1996). A logical DES approach to the design of hybrid systems. *Mathematical and Computer Modeling*, 23(11/12):55–76.

Tacconi, D. A., Lewis, F. L., and Huang, H.-H. (1996). Modeling and simulation of discrete event systems using a matrix formulation. In *Proceedings of the 4th IEEE Mediterranean Symposium on New Directions in Control and Automation*, pages 590–594, Maleme, Krete, Greece.

Valette, R. (1986). Nets in production systems. In *Lecture Notes in Computer Science: Advances in Petri Nets Part II – Petri Nets: Applications and Relations to Other Models of Concurrency*, volume 255, pages 191–217. Springer-Verlag, Berlin; New York.

Valette, R., Courvoisier, M., Demmou, H., Bigou, J. M., and Desclaux, C. (1985). Putting Petri nets to work for controlling flexible manufacturing systems. In *Proceedings of ISCAS '85*, pages 929–932, Kyoto, Japan.

Wonham, W. M. and Ramadge, P. J. G. (1987). On the supremal controllable sublanguage of a given language. *SIAM Journal of Control Optimization*, 25(3):637–659.

Xing, K.-Y., Hu, B.-S., and Chen, H.-X. (1996). Deadlock avoidance policy for Petri net modeling of flexible manufacturing systems with shared resources. *IEEE Transactions on Automatic Control*, 41(2):289–295.

Yamalidou, E. (1991). *Modeling, Optimization and Control of Discrete-Event Chemical Processes using Petri Net Theory*. PhD thesis, Department of Chemical Engineering, University of Notre Dame, Notre Dame, IN.

Yamalidou, K. and Kantor, J. C. (1991). Modeling and optimal control of discrete-event chemical processes using Petri nets. *Computers in Chemical Engineering*, 15(7):503–519.

Yamalidou, K., Moody, J. O., Lemmon, M. D., and Antsaklis, P. J. (1994). Feedback control of Petri nets based on place invariants. Technical Report of the ISIS Group ISIS-94-002.2, University of Notre Dame, Notre Dame, IN.

Yamalidou, K., Moody, J. O., Lemmon, M. D., and Antsaklis, P. J. (1996). Feedback control of Petri nets based on place invariants. *Automatica*, 32(1):15–28.

Zhou, M. C. and DiCesare, F. (1989). Adaptive design of Petri net controllers for error recovery in automated manufacturing systems. *IEEE Transactions on Systems, Man, and Cybernetics*, 19(5):963–973.

Zhou, M. C. and DiCesare, F. (1993). *Petri Net Synthesis for Discrete Event Control of Manufacturing Systems*. Kluwer Academic Publishing, Norwell, MA.

List of Symbols

$\|x\|$ The support of x, i.e., a set of PN places corresponding to the nonzero elements of the vector x.

$\leq$ When the less-than-or-equal-to symbol is applied to vectors or matrices, as in $A \leq B$, it means that every element in A is less than or equal to the corresponding element in B.

$\rightarrow$ Logical implication. $a \rightarrow b$: if a is true then b is true. The symbol is also used to represent a Petri net arc.

$\Rightarrow$ Transformation. $a \Rightarrow b$: a is transformed to b.

$\Leftarrow$ Becomes. $a \Leftarrow b$: a becomes b.

b The right hand side of a constraint inequality. $L\mu_p \leq b, b \in \mathbb{Z}^{n_c}$ or $l^T \mu_p \leq b, b \in \mathbb{Z}$.

c A place in a Petri net (PN) controller.

D A general PN incidence matrix, or the incidence matrix of the closed loop plant/controller system $D = \begin{bmatrix} D_p \\ D_c \end{bmatrix}$.

D_c The incidence matrix of a PN controller. $D \in \mathbb{Z}^{n_c \times m}$.

D_p The incidence matrix of a PN plant. $D \in \mathbb{Z}^{n \times m}$.

D_{uc} The incidence matrix of the uncontrollable portion of a PN plant. D_{uc} is composed of the columns of D_p corresponding to uncontrollable transitions. $D_{uc} \in \mathbb{Z}^{n \times n_{uc}}$.

D_{uo} The incidence matrix of the unobservable portion of a PN plant. D_{uo} is composed of the columns of D_p corresponding to unobservable transitions. $D_{uo} \in \mathbb{Z}^{n \times n_{uo}}$.

I An identity matrix.

l Parameter in a constraint inequality, $l^T \mu_p \leq b, l \in \mathbb{Z}^n$.

L Parameter in a constraint inequality, $L\mu_p \leq b, L \in \mathbb{Z}^{n_c \times n}$.

μ The marking vector of a general PN, or the marking of the closed loop plant/controller system.

μ_0 The initial marking of the PN with marking vector μ.

μ_c The marking vector of a PN controller. $\mu_c \in \mathbb{Z}^{n_c}$.

μ_{c_0} The initial marking of a PN controller.

μ_p The marking vector of a PN plant. $\mu_p \in \mathbb{Z}^n$.

μ_{p_0} The initial marking of a PN plant.

m The number of transitions in plant Petri net.

n The number of places in the plant Petri net.

n_{uc} The number of uncontrollable transitions in a PN plant.

n_{uo} The number of unobservable transitions in a PN plant.

n_c The number of individual vector constraints in $L\mu_p \leq b$.

p A PN place.

P The set of places for a Petri net.

$\bullet p$ The set of transitions with output arcs to place p.

$p\bullet$ The set of transitions receiving input arcs from place p.

q The firing vector of a PN. $q \in \mathbb{Z}^m$.

s A nonnegative integer vector whose support is a trap or siphon. $s \in \mathbb{Z}^n$.

S A set of PN places, usually referring to a trap or siphon.

$\bullet S$ The set of transitions with output arcs to all places in the set S.

$S\bullet$ The set of transitions receiving input arcs from all places in the set S.

t A PN transition.

$\bullet t$ The set of places with output arcs to transition t.

$t\bullet$ The set of places receiving input arcs from transition t.

T Superscript for vector or matrix transpose.

T The set of transitions for a Petri net.

T_c The set of controllable transitions in a PN plant.

T_{uc} The set of uncontrollable transitions in a PN plant. $T_c \cap T_{uc} = \emptyset$.

$_{uc}$ Subscript referring to uncontrollability.

$_{uo}$ Subscript referring to unobservability.

x A PN place invariant vector. $x \in \mathbb{Z}^n$.

X A matrix, the columns of which form a basis for a PN's place invariants. If $x^T D = 0$, then x is linearly dependent with the columns of X.

y A PN transition invariant vector. $y \in \mathbb{Z}^m$.

Y A matrix, the columns of which form a basis for a PN's transition invariants. If $Dy = 0$, then y is linearly dependent with the columns of Y.

$\mathbb{Z}$ The set of integers.

$\mathbb{Z}^n$ The set of integer vectors with n elements (dimensions).

$\mathbb{Z}^{n \times m}$ The set of integer matrices with n rows and m columns.

Index

Italicized page numbers indicate definitions within the text. An 'n' is appended to a page number when the entry refers to a footnote; 'g' refers to glossary entries.

About the Authors

John Moody received his doctorate in electrical engineering in 1998 from the University of Notre Dame, where he also earned the B.S. and M.S. degrees.

His research interests include discrete event and hybrid control systems, neural network construction and architecture, parallel computing, and autonomous, intelligent control systems. He has authored a number of publications on these subjects in journals and conference proceedings.

During his undergraduate education, he worked as an engineering intern for the Bendix Engine Controls Division of the Allied Signal Aerospace Company. He has been awarded an Arthur J. Schmitt fellowship and the Center for Applied Mathematics fellowship from the University of Notre Dame. He is a member of the Eta Kappa Nu and Tau Beta Pi honor societies and the IEEE.

Panos J. Antsaklis is Professor of Electrical Engineering at the University of Notre Dame. He received his undergraduate degree from the National Technical University of Athens (NTUA), Greece, and his M.S. and Ph.D. degrees in electrical engineering from Brown University. He has taught and conducted research at Rice University, Imperial College of the University of London, MIT, NTUA and the Technical University of Crete, Greece.

His research interests are in the area of systems and control, with emphasis on hybrid and discrete event systems, on autonomous, intelligent and learning control systems, on methodologies for reconfigurable control and neural networks. He has authored a number of publications in journals, conference proceedings and books, and he has edited three books: *An Introduction to Intelligent and Autonomous Control* (Kluwer Academic 1993; with K. Passino), *Hybrid Systems II* and *Hybrid Systems IV* (Springer 1995 and 1997; with W. Kohn, A. Nerode and S. Sastry). He has also authored a graduate textbook *Linear Systems* (McGraw-Hill 1997; with A.N. Michel).

He serves in the editorial boards of several journals, he has been the guest editor of special issues on neural networks (*IEEE Control Systems magazine*; 1990 and 1992), on intelligence and learning (*IEEE Control Systems magazine*; 1995), on hybrid control systems (*IEEE Transactions on Automatic Control* 1998) and on hybrid systems (*Journal of Discrete Event Dynamic Systems*, 1998). He has served as program chair and general chair of major systems and control conferences and he was the 1997 President of the IEEE Control Systems Society (CSS). He is an IEEE Fellow.